GROUND PENETRATING RADAR INVESTIGATIONS IN UPPER KAMA

POTASH MINES

by

OLEG NIKOLAIEVICH KOVIN

A DISSERTATION

Presented to the Faculty of the Graduate School of the

MISSOURI UNIVERSITY OF SCIENCE AND TECHNOLOGY

In Partial Fulfillment of the Requirements for the Degree

DOCTOR OF PHILOSOPHY

in

GEOLOGY AND GEOPHYSICS

2010

Approved by

Neil L. Anderson, Advisor
Estella Atekwana
David J. Rogers
Derek Apel
Alexey Malovichko

UMI Number: 3462712

UMI 3462712

ProQuest LLC.
789 East Eisenhower Parkway
P.O. Box 1346
Ann Arbor, MI 48106 - 1346

## PUBLICATION DISSERTATION OPTION

This dissertation has been prepared in the style utilized by the Journal of Applied Geophysics. Pages 104 – 141 will be submitted for publication in that journal.

# ABSTRACT

An understanding of the structure and state of the rock mass surrounding underground openings in the potash mines is critically important for safe mining, planning the methods of extraction of an orebody, and preventing the influx of ground water.

Continuous common offset ground penetrating radar (GPR) data were acquired in the potash mine operated by the Joint Stock Company (JSC) "Silvinit" (Russia) as part of an investigation of both pre-existing fractures exposed by mine workings and other anomalous geological structures.

During the course of GPR investigation, the electrical properties of salt-bearing units were determined, site-specific data acquisition techniques and object-oriented data processing schemes adapted to the geological and geotechnical environment of the Upper Kama potash deposit were developed, and the methodology of 2-D and 3-D GPR data interpretation using interactive modeling was worked out.

Open fractures and fault and fold features were successfully mapped using 2-D and 3-D GPR techniques. FK filtering significantly improved the reliability of fracture detection. Spatial models of mapped fractures were created using 3-D GPR imaging technique. Migration of the georadar data was required to obtain the true geometry of folded salt beds.

The results of this GPR-based investigation demonstrate that the ground penetrating radar georadar method is capable of providing valuable information about deformation structures within the evaporite units of the Upper Kama potash deposit.

# ACKNOWLEDGMENTS

I would like to thank my advisor, Dr. Neil Anderson, and committee members, Dr. Estella Atekwana, Dr. David Rogers, Dr. Derek Apel, and Dr. Alexey Malovichko for their valuable suggestions, flexibility in solving my problems and careful editing. I would like to thank the faculties and staff of the Department of Geological Sciences and Engineering who made my stay at the university fruitful and enjoyable. Especially, I'm grateful to "Mom" of all the department students, Katherine Mattison, and Paula Cochran who kindly helped me to manage the hard routine paper work. I would like also to thank the department chair, Dr. Robert Laudon, and Drs. Francisca Oboh-Ikuenobe, John Hogan, Stephen Gao, and Norbert Maerz for interesting discussions and assistance in gaining new experiences. I also thank to former graduate students Moidaki Moikwathai, Niklas Putnam, Ahmed Ismail, and Thanop Thitimakorn for collaborating in solving many scientific problems and creating a warmest atmosphere in the office. I thank the Department of Geological Sciences and Engineering for supporting fieldwork in Russia during the summer semester. I also thank the JSC "Silvinit" and senior geologist, Yuri Mynka, for permitting and assisting georadar investigations in their potash mine. I thank my colleagues from the Mining Institute UB RAS and, especially, Dr. Alexey Kudryashov for their assistance with my experiments and for providing necessary geological information.

Finally, I am very grateful to my family for incredible patience, support and understanding.

I respectfully dedicate this work to the memory of my mother who passed away a year ago.

# TABLE OF CONTENTS

# LIST OF ILLUSTRATIONS

# LIST OF TABLES

# NOMENCLATURE

Symbol          Description

# 1. INTRODUCTION

## 1.1. PURPOSES OF WRITING THIS DISSERTATION

Gradual and catastrophic surface subsidence in response to the closure of anthropogenic openings within potash mines and/or the influx of overlying ground water through open fractures can cause serious damage to property, infrastructure and industrial facilities. For example, significant damage to man-made surface structures was reported after the catastrophic collapses at the Berezniki potash mine 3 in 1988 and Berezniki potash mine 1 in 2006 (Belkin, 2010). Mined zones within the Upper Kama potash deposit underlie parts of the cities of Berezniki and Solikamsk (Perm kray, Russia). These areas are very high risk because of population and infrastructure densities. Chemical and power plants, roadways, railroads and gas pipelines located within the subsidence zone are only a few of the features at risk.

The stability of the underground openings within the Perm region potash mines depends on the mining methods employed, the geomechanical properties of the pillars supporting the overlying beds, the integrity and structure of the overlying impermeable strata, and the subsidence prevention measures employed.

This author has chosen to investigate man-made and pre-existing deformation features within the rock mass surrounding mine openings because the understanding of the integrity of mine openings is critically important for safe mining and the prevention of catastrophic ground water inflow. The author has elected to use the ground penetrating radar (GPR) tool to investigate mine openings because this technology is capable of providing high-resolution images of fractures and folds with rock salt.

## 1.2. GPR AS A HAZARD RISK EVALUATION METHOD

Only limited exploratory drilling (boreholes) is permitted in the Perm region potash mines because of concerns that integrity of the overlying impermeable (protective) strata could be compromised. Therefore, non-invasive geophysical methods are widely used to map strata and structures within and above/below the mined zones and identify/investigate potentially hazardous features (Chouteau et al., 1997; Eso et al., 2006; Gendzwill, 1969; Gendzwill and Stead, 1992; Neal et al., 1995; Thoma et al., 2003; Yaramanci, 2000).

GPR (referred to herein as georadar, and/or ground penetrating radar) is an effective geophysical imaging tool, with a wide set of applications to geological mapping and underground mining (Annan, 2002). The successful application GPR to fracture detection and study of deformation in crystalline and sedimentary rocks is reported in numerous publications (Grandjean, 1996; Busby, 1999; Seol et al., 2001; Haeni et al., 2002; Orlando 2003; Porsani et al., 2006). The detection and mapping the millimetric cracks using GPR are described by Toshioka et al. (1995), Grasmueck (1996), Lane et al. (2000), and Laurence et al., 2003. A number of studies have demonstrated the enhanced capability of 3-D GPR imaging to the delineation of the subsurface structures (Grasmueck et al., 2004; Gross et al., 2002, 2003, 2004; Christie et al., 2009).

There is less publically-available information regarding the specific application of GPR to underground mining. Cook (1969, 1975) and Coon et al. (1981) describe some of the earliest reported GPR experiments in coal and salt mines. The first experiments conducted in salt mines demonstrated that high-frequency electromagnetic waves are able to penetrate substantial distances in salt because of the relative uniformity and low conductivity of these strata (Cook, 1969; Holzer at al., 1972; Stewart and Unterberger, 1976; Unterberger, 1978). The GPR method has been successfully used in salt mines by this author and others to map stratigraphy, estimate the thickness of the overlying water-protective beds, characterize fractures, detect unstable roof rock, and evaluate the integrity of supporting pillars (Kovin, 2002; Thoma et al., 2003; Maybee et al., 2004). Annan et al. (1988), Gregoire and Halleux (2002), and Kovin (2010) describe the successful detection of fractures in potash mines.

Preliminary experiments by the author and colleagues in the Upper Kama potash mines demonstrated that the GPR method is capable of providing detailed and effectively lineally-continuous information about rock mass structures at distances of up to 40 m from mine openings (Kovin et al., 2002). The non-destructive, portable and cost-effective method, georadar is well suited for usage in the salt mine environment. Although the GPR method has been extensively used in German and Canadian salt mines for tens of years (Band et al., 1988; Annan et al., 1988; Chouteau et al., 1997), studies have shown that the GPR acquisition, data processing, and interpretation methodologies require an adaptation to local geological and mining environment.

## 1.3. OBJECTIVES AND APPROACH

This investigation has the following specific objectives:

- to determine the electrical properties of the salt rocks[1] of the Upper Kama potash deposit;

---

[1] The terms "salt rock" and "salt rocks", as well as "salt", "salts", are used in this dissertation to refer to any lithotype of evaporitic rock comprised of minerals sylvite, carnallite and halite or their mix.

- to develop site-specific acquisition technique including consideration of antennas directionality and geometry of mine workings space;
- to evaluate the effectiveness of processing procedures and parameters of processing flow, and to develop an object-oriented data processing schemes adapted to the local geological and geotechnical environment;
- to develop a methodology of GPR data interpretation for the specific objects of investigation.

About 5 km of continuous common offset GPR data (mostly multiple separate 2-D profiles) were collected in the potash mine of the JSC "Silvinit" located near the city of Solikamsk (Russia) as part of an effort to develop data acquisition, processing and interpretation methodologies specifically adapted to the Upper Kama potash deposit. GPR data were collected in the upper mining level where sets of tension vertical fractures were exposed by mine workings as well as in a conveyor drift within the underlying rock salt (in an effort to investigate deeper geological structures and features beneath the mine openings).

A small subset of GPR data (seven parallel 8 m long profiles spaced at 0.2 m) was used to create a 3-D grid for imaging, on the millimetric scale, a surface-exposed fracture.

The OKO (Logical Systems, Russia) was employed for data acquisition. Reflexw (Windows OS) and OpendTect (Linux OS) software were used for data processing and the interpretation of 2-D and 3-D data, respectively.

The processing sequence was subdivided into:

- Common: A common processing flow, defined by specifics of the GPR records, was applied to all the GPR data (start time correction, "dewow" filtering, background or "ringing" noise removal, gain correction);

- Object-oriented: Object-oriented processing was also applied to the data. Such processing is dependent on the geometry and physical properties of the target object, the characteristics of the wave field, and physical parameters of media (which must be determined through processing). This processing includes a variety of procedures such as FK filtering, band pass filtering, migration, velocity analysis, etc. For example, because of intensive folding, the GPR sections of the evaporites are characterized by diffractions. Migration is required to make the data visually interpretable.

The results of data analysis show that GPR method can be of utility in terms of the detection of small-scale fractures, folds of various ranges, and fault structures in salt rocks at the depth of up to 20 m. The 3-D GPR imaging technique proved to be an effective tool for investigating the geometry of fractures.

This dissertation research has the following results: the electrical properties of typical evaporite formation members were determined; site-specific data acquisition technique and object-oriented data processing schemes based on the analysis of effectiveness of different processing procedures and processing parameters were developed; and methodologies for 2-D and 3-D GPR data interpretation using interactive modeling were designed.

## 1.4. STRUCTURE OF THE DISSERTATION

Electromagnetic waves theory as it relates to GPR method is presented in Section 2. The geological and mining settings of the Upper Kama potash deposit are described in Section 3 along with a review of the main problems associated with mining of the salt rocks. Section 4 is a brief summary of experimental work setup and velocity determination procedures. Section 5 is a description of the results of experimental studies of fractures, fold and fault structures within salt rocks and will be submitted to the Journal of Applied Geophysics as stand-alone article.

## 2. GEOLOGICAL AND MINING SETTING

### 2.1. GENERAL LOCATION

The Upper Kama (Verkhnekamskoye) potash deposit (also called Upper Kama salt deposit) is located in the northern part of the Perm kray (Russia), about 250 km north of the city of Perm (Figure 2.1). Evaporated sodium chloride salt has been produced from the brine in this area since the beginning of 15th century. The first salt boreholes and evaporation facilities appeared near the present-day city of Solikamsk. The potash salts were discovered in 1925 during the drilling of an oil exploration borehole. Now, the Joint Stock Company (JSC) "Uralkali", headquartered in city of Berezniki, operates the mines in the southern part of the potash deposit, and JSC "Silvinit", headquartered in Solikamsk, operates the mines in the northern area.

The Upper Kama deposit is second in size to the Saskatchewan Prairie Evaporite (Canada) among the world's currently mined potash occurrences (Garrett, 1995). The Upper Kama potash deposit covers an area of 3 800 km$^2$. It is approximately 135 km long and 40-45 km wide (Kopnin, 1995). Sylvinite beds are located at depths of 75-450 m (Garrett, 1995).

Currently, potash ore is mined at five underground mines, i.e. Solikamsk potash mine 1, Solikamsk potash mine 2, Solikamsk potash mine 3, Berezniki potash mine 2, and Berezniki potash mine 4. Berezniki potash mine 3 was flooded in 1988; the oldest mine operated by JSC "Uralkali", Berezniki potash mine 1, was flooded in 2006. A map showing the Upper Kama potash mines location is presented as Figure 2.2.

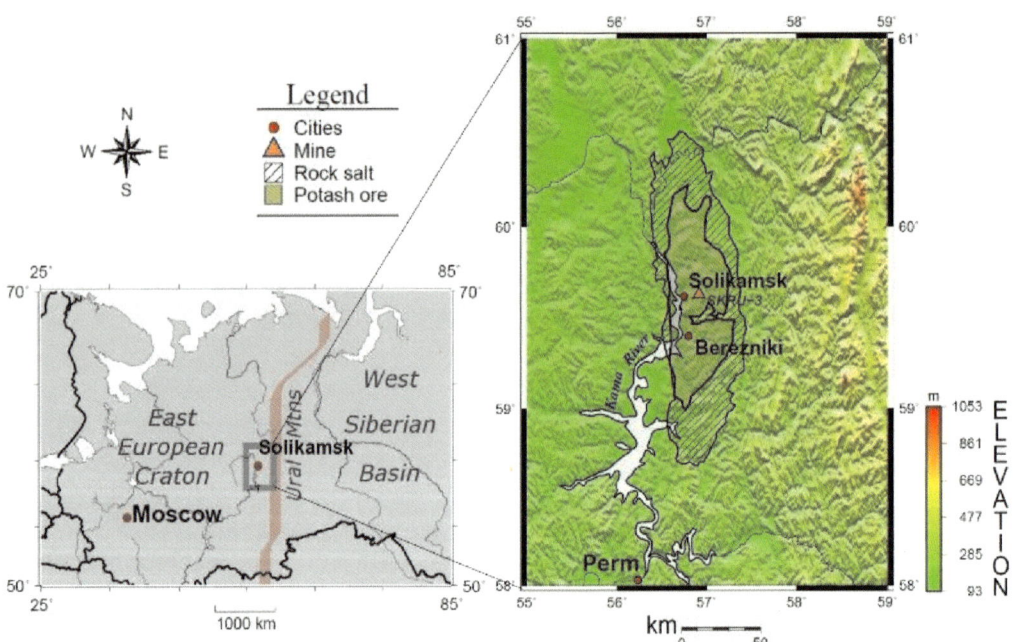

Figure 2.1. Location map of the Upper Kama (Verkhnekamskoye) potash deposit.

Figure 2.2. Location map of the Upper Kama potash mines. Names of the mines are abbreviated in Latin alphabet (BKRU and SKRU correspond to Berezniki potash mine and Solikamsk potash mine, respectively).

## 2.2. GEOLOGY OF THE SOLIKAMSK DEPRESSION

The Upper Kama potash deposit lies within the central part of the Solikamsk (Solikamskaya) depression of the Pre-Ural foreland basin. The Solikamsk depression is bordered to the west by East European Craton and to the east by West Uralian Folding Zone. Location of the Solikamsk depression and the Upper Kama potash deposit is shown in Figure 2.3.

The Permian age (Kungurian Stage) evaporitic formation, a layered sequence of up to 550 m thick, is composed predominantly of halite with substantial interbedded sylvinite, carnallite, clay, and anhydrite beds. The Upper Kama potash deposit consists of potentially mineable potash-bearing beds lying at the top of evaporite strata.

The salts were deposited after the rise of the Ural Mountains in the late Carboniferous. The salt deposit formed via minerals precipitation under solar evaporation processes at shallow restricted basin. A basin was created as a result of the continental collision between the East-European platform and West-Siberian plate (Rodgers, 1990; Zonenshain et al., 1990; Friberg et al., 2002; Warren, 2010). The typical scheme of collision basin evaporite deposition is depicted in Figure 2.4.

Predominantly, sediments outcropping in the Solikamsk depression are Permian in age. The regional geological map of the Solikamsk depression and surrounding area is shown in Figures 2.5a and 2.5b.

**2.2.1. Stratigraphy.** The Permian evaporite formation is underlain by a succession of Proterozoic (Riphean and Vendian Stages) terrigeneous and carbonate-terrigeneous rocks and Paleozoic predominantly carbonate rocks with total thickness of about 4 km. The Paleozoic Devonian, Carboniferous, Permian sediments disconformably overlie the Proterozoic complex (Figure 2.6).

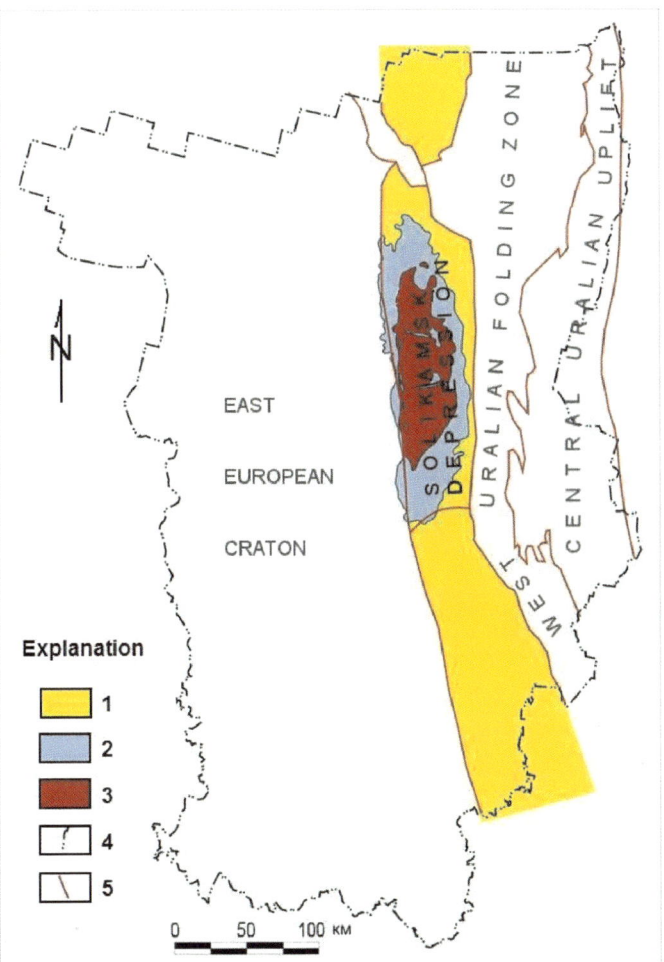

Figure 2.3. Generalized tectonic scheme of the Permskiy kray and location of the Solikamsk depression (modified from Chaykovskiy et al, 2009). Explanation: 1 – Pre-Uralian foreland basin; 2 – rock salt[2] distribution; 3 – potash salts[3] distribution; 4 – administrative border; 5 – boundaries of main tectonic structures.

---

[2] The term "rock salt" is used to refer to evaporitic rock predominantly consisted of mineral halite.

[3] The term "potash salts" is used to refer to evaporitic rock consisted of minerals sylvite and carnallite.

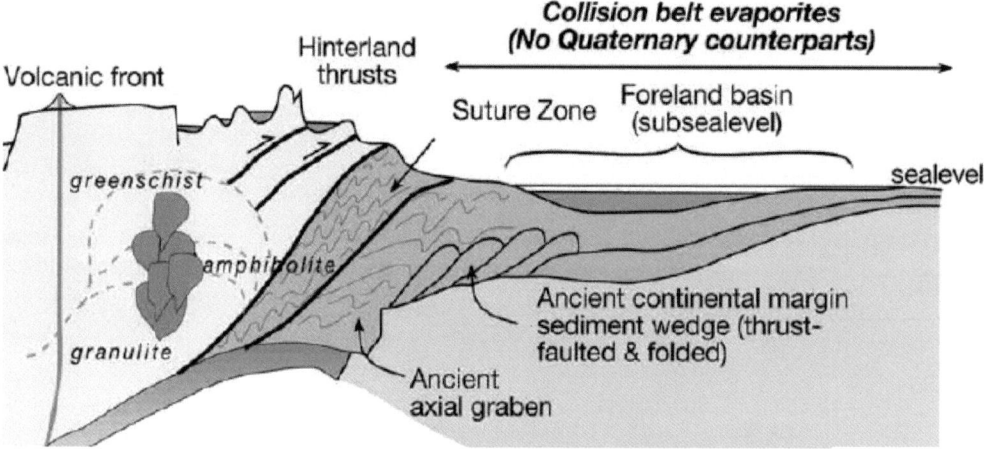

Figure 2.4. Generalized scheme of evaporites deposition in a collision basin

environments (modified from Warren, 2010).

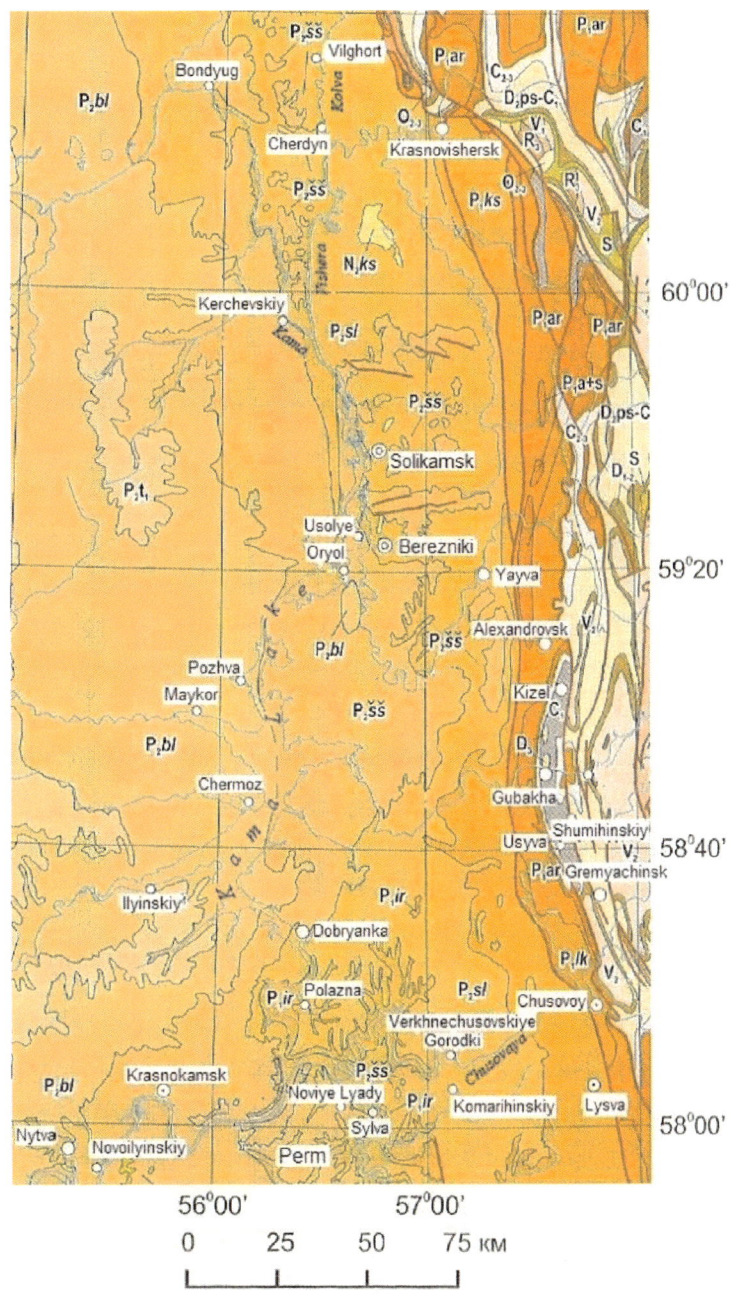

Figure 2.5. a. Regional geological map of the Solikamsk depression and surrounding area
(modified from Yechlakov and Morozov, 2006).

## LEGEND

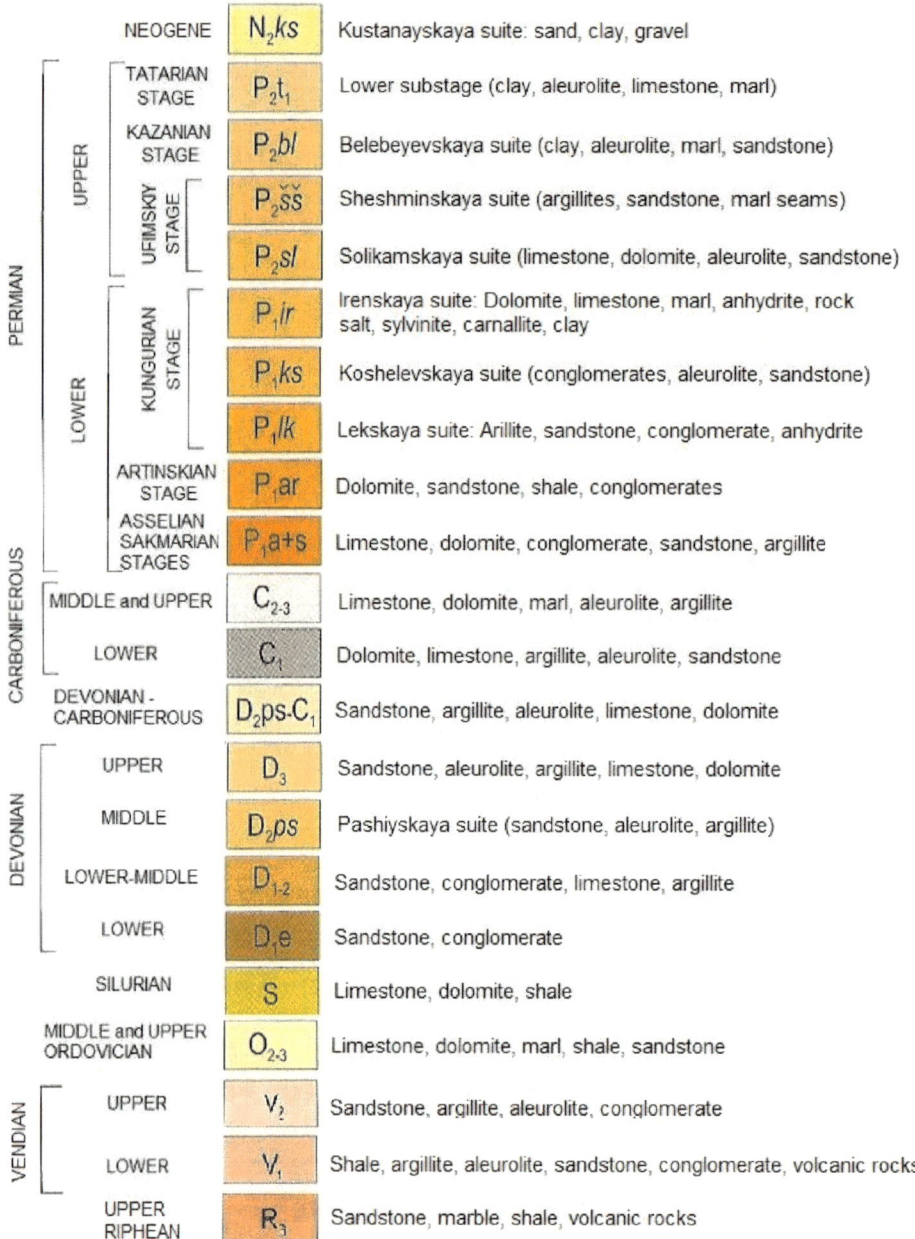

| | | | |
|---|---|---|---|
| NEOGENE | | $N_2ks$ | Kustanayskaya suite: sand, clay, gravel |

PERMAN — UPPER — TATARIAN STAGE — $P_2t_1$ — Lower substage (clay, aleurolite, limestone, marl)

KAZANIAN STAGE — $P_2bl$ — Belebeyevskaya suite (clay, aleurolite, marl, sandstone)

UFIMSKIY STAGE — $P_2\check{s}\check{s}$ — Sheshminskaya suite (argillites, sandstone, marl seams)

$P_2sl$ — Solikamskaya suite (limestone, dolomite, aleurolite, sandstone)

LOWER — KUNGURIAN STAGE — $P_1ir$ — Irenskaya suite: Dolomite, limestone, marl, anhydrite, rock salt, sylvinite, carnallite, clay

$P_1ks$ — Koshelevskaya suite (conglomerates, aleurolite, sandstone)

$P_1lk$ — Lekskaya suite: Arillite, sandstone, conglomerate, anhydrite

ARTINSKIAN STAGE — $P_1ar$ — Dolomite, sandstone, shale, conglomerates

ASSELIAN SAKMARIAN STAGES — $P_1a+s$ — Limestone, dolomite, conglomerate, sandstone, argillite

CARBONIFEROUS — MIDDLE and UPPER — $C_{2-3}$ — Limestone, dolomite, marl, aleurolite, argillite

LOWER — $C_1$ — Dolomite, limestone, argillite, aleurolite, sandstone

DEVONIAN - CARBONIFEROUS — $D_2ps-C_1$ — Sandstone, argillite, aleurolite, limestone, dolomite

DEVONIAN — UPPER — $D_3$ — Sandstone, aleurolite, argillite, limestone, dolomite

MIDDLE — $D_2ps$ — Pashiyskaya suite (sandstone, aleurolite, argillite)

LOWER-MIDDLE — $D_{1-2}$ — Sandstone, conglomerate, limestone, argillite

LOWER — $D_1e$ — Sandstone, conglomerate

SILURIAN — $S$ — Limestone, dolomite, shale

MIDDLE and UPPER ORDOVICIAN — $O_{2-3}$ — Limestone, dolomite, marl, shale, sandstone

VENDIAN — UPPER — $V_2$ — Sandstone, argillite, aleurolite, conglomerate

LOWER — $V_1$ — Shale, argillite, aleurolite, sandstone, conglomerate, volcanic rocks

UPPER RIPHEAN — $R_3$ — Sandstone, marble, shale, volcanic rocks

Figure 2.5. (Continued) b. Legend for regional geological map.

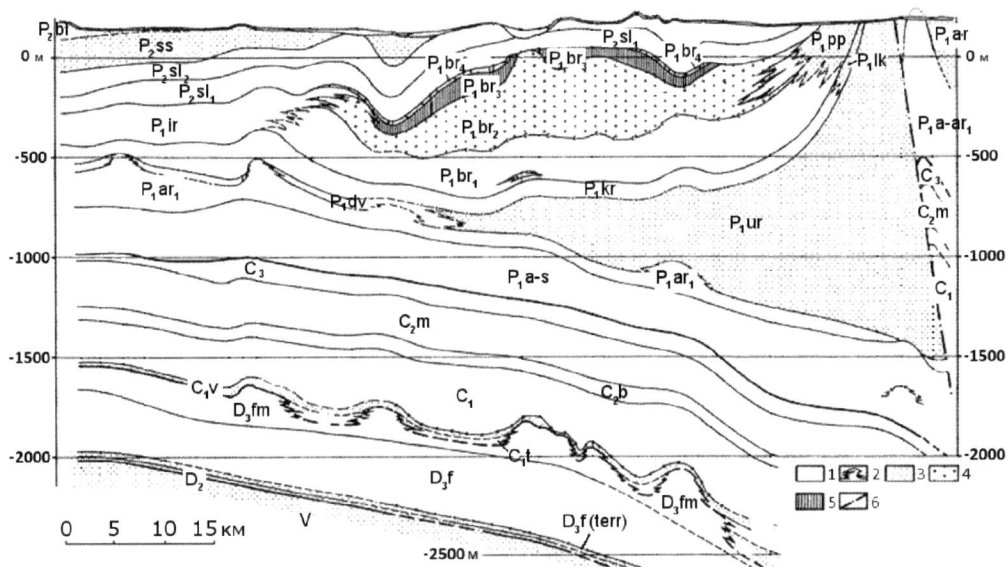

Figure 2.6. W-E geological cross section of Solikamsk depression (modified from Kudryashov, 2001). Explanation: 1 – dominantly carbonate sediments; 2 – reef structures; 3 – dominantly terrigeneous sediments; 4 – rock salt; 5 – potash salts; 6 – Vsevolodo-Vilvenskiy thrust fault.

**2.2.1.1 Devonian System.** Middle and Upper Devonian Series are represented by terrigeneous and carbonate members. Terrigeneous member consists of aleurolite, sandstone and argillites of Middle Devonian Series (Eifelian, Givetian Stages) and lower part of Upper Devonian (Frasnian Stage $D_3f(terr)$). The carbonate Upper Devonian is composed of shallow-water biogenic-carbonate buildup deposits of the Frasnian ($D_3f$) and Famennian ($D_3fm$) Stages.

**2.2.1.2 Carboniferous System.** The Lower Series include the clayey limestone, dolomite, argillites, and organic-reach sandstone of shallow-marine Tournaisian ($C_1t$) and Visean ($C_1v$) sediments. The Middle Carboniferous is represented by detrital biogenic limestone, dolomite and argillites of Bashkirian ($C_2b$) and Moscovian ($C_2m$) petroliferous sediments. The Upper Carboniferous ($C_3$) is composed dominantly of dolomite.

**2.2.1.3 Permian System.** Permian System in the study area includes the Lower and Upper Series. The Asselian and Sakmarian ($P_1a+s$) Stages of Lower Permian Series are represented by limestone that is locally petroliferous. Artinskian ($P_1ar$) sediments are comprised of terrigeneous and carbonate strata. Carbonate sediments ($P_1ar_1$) are represented by the biogenic detrital limestone and, locally, develop low amplitude reef structures. The terrigeneous sediments occur in the eastern part of Solikamsk depression, where it forms flysch-molasse "wedge" referred to as Urminskaya suite ($P_1ur$).

Argillites, aleurolite, sandstone, and conglomerates represent the molasse sediments. The thickness of Permian molasse "wedge" increases eastward from 120 m to above 1500 m. The molasse sediments grade westward to clayey limestone, marl and dolomite of the Divyinskaya suite ($P_1dv$).

The Lower Permian Kungurian Stage includes sediments of Irenian ($P_1ir$) suite, Karnaukhovskaya suite ($P_1kr$), and Lekskaya suite ($P_1lk$). Irenian limestone, dolomite, and anhydrite beds occur only in the western part of the Solikamsk depression. Carbonate-sulphate sediments of Karnaukhovskaya suite (Figure 2.6) are located in the central part of depression and grade eastward to clastic rocks of the Lekskaya suite (Figure 2.5a). Bereznikovskaya suite ($P_1br$) occurs in the central part of the Solikamsk depression and is presented by clay-anhydrite and evaporite complexes. Popovskaya suite ($P_1pp$) occurs in the eastern part of Pre-Uralian foreland basin. It is comprised of the marl, clay layers, and the anhydrite and salt lenses.

The Upper Permian Series include Ufimian and Kazanian Stages ($P_2kz$). Solikamsk Horizon (suite) of the Ufimian Stage is subdivided to salt-marl subformation (Niznesolikamskaya subsuite $P_2sl_1$) and terrigeneous-carbonate subformation ($P_2sl_2$). Sheshma Horizon ($P_2ss$) of the Ufimskiy Stage overlying Solikamsk suite consists dominantly of limy sandstone and aleurolite (Lozovsky et al., 2009). Kazanian sediments occur to the west from the salt deposit and also comprise dominantly of sandstone and aleurolite.

**2.2.1.4 Evaporite formation.** Evaporite formation of the Solikamsk depression includes evaporitic sediments of Karnaukhovskaya and Bereznikovskaya suites, and Nizhnesolikamskaya subsuite ($P_2sl_1$). The local stratigraphic and litho-stratigraphic columns of the evaporite formation used at the Upper Kama potash deposit is shown in Figure 2.7.

Karnaukhovskaya suite is represented by four interbedded series of sulphate and carbonate rocks of total thickness of about 100 m.

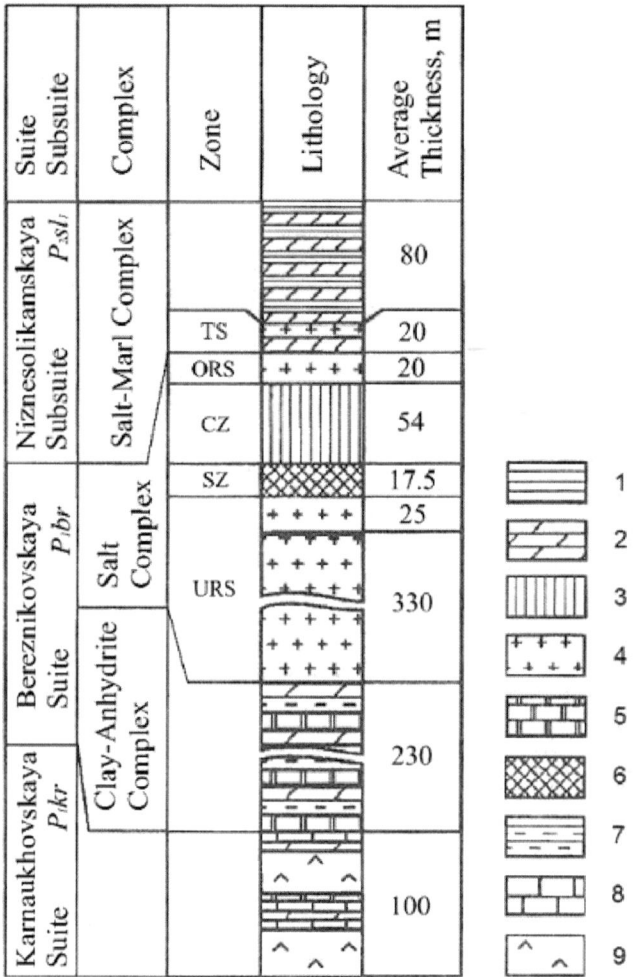

Figure 2.7. Local stratigraphic and litho-stratigraphic columns of the evaporite formation of the Solikamsk depression (modified from Kudryashov, 2001). Explanation: 1 – clay; 2 – marl; 3 – rock salt; 4 – carnallite and rock salt; 5 – sylvinite and rock salt; 6 – argillite; 7 – dolomite; 8 – limestone; 9 – anhydrite. The members of upper part of evaporite formation are indicated as: URS – underlying rock salt; SZ – sylvinite zone; CZ-carnallite zone; ORS – overlying rock salt; TS – transition sequence.

Bereznikovskaya suite is subdivided into the *clay-anhydrite complex* (CAC) and *salt complex* (SC). Clay-anhydrite complex ($P_1br_1$) is deposited in early stage of forming the evaporite basin and is comprised of marl, argillites and of lesser amount of anhydrite, limestone, rock salt, aleurolite, and sandstone. Average thickness of this complex is about 230 m. Salt complex of total thickness of up to 550 m is subdivided to the *underlying rock salt* (URS) ($P_1br_2$), *potash deposit* ($P_1br_3$) and *overlying rock salt* (ORS) ($P_1br_4$).

Salts of evaporite formation of the Solikamsk depression are represented by rock salt, sylvinite and carnallite. The term potash denotes a variety of mined and manufactured salts, all containing the element potassium in water-soluble form. Three minerals: sylvite (potassium chloride - KCl), carnallite (hydrated potassium magnesium chloride – $KMgCl_3 6H_2O$) and halite (sodium chloride – NaCl) with subordinated amounts of anhydrite and clay form a potash ore. Potash salts indicate the final stage of formation of the evaporite sequence.

Sylvinite, as the most sylvite-rich crude ore, is mined in the Upper Kama deposit. There are three distinguished types of sylvinite ore (referred to as *red*, *banded* and *multi-colored* sylvinite), which differ in texture and sylvite content.

The sequence of salts, anhydrite and clayey carbonates of the underlying rock salt of up to 500 m thick lies in the base of evaporites.

The upper part of salt complex of 80-120 m thick contains economic deposits of potash salts, which include 13 sylvite and carnallite mineable beds and halite interlayers (Garrett, 1995, Kudryashov, 2001). Four sylvite-rich members (referred to as A, Red I (RI), Red II (RII) and Red III (RIII)) in the lowest part of productive strata are the principal mineable beds (Figure 2.8).

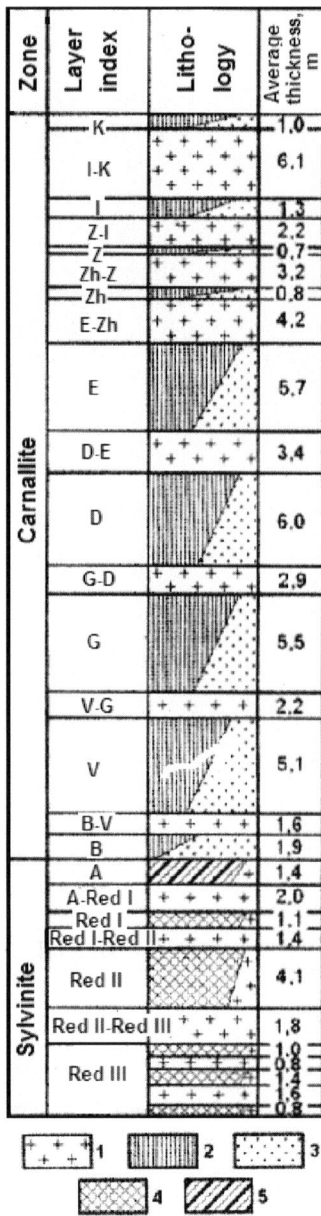

Figure 2.8. Litho-stratigraphic column of the potash members (modified from

Kudryashov, 2001). Explanation: 1 – rock salt; 2 – carnallite rock; 3 – multi-colored

sylvinite; 4 – red sylvinite; 5 – banded sylvinite.

The average thickness of a sylvinite zone is 17.4 m ranging from 3.3 to 30 m. Nine carnallite beds (namely B, V, G, D, E, Zh, Z, I, K) along with interbed rock salt named respectively to the adjacent beds lie on top of the *banded* sylvinite layer A and form a carnallite zone that ranges in thickness from 38 to 80 m (Kudryashov, 2001).

Locally, the sylvinite beds are replaced by barren halite and a host rock of carnallite beds is replaced by the *multi-colored* sylvinite or halite. The bonded A and B layers are usually mined together and are referred to as AB bed.

Competent halite layer of 20 m thick, termed *overlying rock salt* (ORS), lies on a top of the upper carnallite bed. However, there are the areas, where the upper salt layer is washed out from the top of uplift structures.

Insolubles are presented in evaporite formation by thin centimeter scale parting seams composed of clay and anhydrite. The 1 - 2 m thick carbonate clay layer (referred to as Marker Clay) located in the underlying rock salt 20 m below sylvinite zone is the most stable stratigraphic marker at the Upper Kama deposit (Kudryashov et al., 2004). The *salt-marl complex* (SMC) represents the Nizhnesolikamskaya subsuite ($P_2sl_1$) lying on the top of salt complex. The average content of complex is: marl and clay – 65%, rock salt – 30%, gypsum and anhydrite – 5% (Kudryashov, 2001). Lower part of the salt-marl complex containing the rock salt beds is termed *transition sequence* and is approximately of 20 m thick. Lithology and gamma-ray log chart corresponding to interval of the transition sequence is shown in Figure 2.9. The low gamma-ray curve values indicate the rock salt layers. The high gamma-ray curve values correspond to clayey rock.

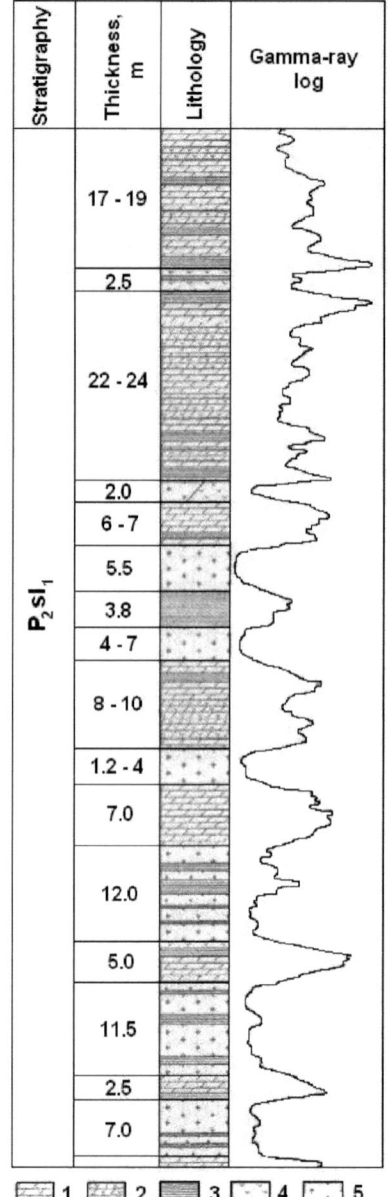

Figure 2.9. Lithostratigraphic column and gamma-ray log profile (out of scale) of salt-marl complex (modified from Kudryashov, 2001). Explanation: 1 – marl; 2 – gypsum-marl; 3 – clay; 4 – gypsum; 5 – rock salt.

The upper part of potash beds along with the overlying marl, halite, clay, anhydrite, gypsum and carbonate beds of the *transition sequence* (TS) is left unmined to form a water protective barrier above the mine. The thickness of impermeable strata varies from 13.5 to 135 m and averages 70–90 m (Polyanina, 1995; Kovin, 1997).

**2.2.1.5 Stratigraphy of overburden sediments.** Limestone, dolomite, marl followed by clastic sediments with multiple aquifer layers of total average thickness of 200 m overlie evaporites. For salt deposits, the knowledge about location and dynamics of groundwater is very important because it determine the methods of excavation and mining safety requirements.

Limestone, marl, dolomite, and overlying sandstone, aleurolite, and argillite of upper subdivision of Solikamsk Horizon ($P_2sl_2$) form the *terrigeneous-carbonate complex* (TCC) of 90 to 170 m thick that is followed by sandstone and aleurolite sediments of Sheshma Horizon ($P_2ss$) referred to as *multi-colored complex* (MCC). The thickness of *multi-colored complex* at deposit area ranges from 0 to 675 m (Kudryashov, 2001).

**2.2.2. Evaporites Tectonics.** Tectonics of the Upper Kama potash deposit is defined by the structure of underlying sedimentary and crystalline strata, inner tectonics of salt formation, tectonic peculiarities of overlying sediments and geodynamic regime of the area. Predominantly, tectonic movements in the adjacent orogens govern the deformation of the evaporites formations (Jeremic, 1994). The Solikamsk depression is located in the central part of Pre-Uralian foreland basin that constituted the conditions of deposition and deformation of evaporite rocks.

**2.2.2.1 Regional tectonics.** The crystalline basement tectonics plays leading role in evolution of overlying sediments. The crystalline basement has not been explored with boreholes on the area of the Solikamsk depression. The principal information about basement structure is obtained from geophysical data. Results of the seismic and gravity investigations confirm the gently eastward dipping surface of crystalline rock and reveal the main elements of regional fault system (Kudryashov, 2001). General structure of basement of the Solikamsk depression is shown in Figure 2.10. The overlying sediments succeed a dominant submeridianal orientation of basement structures.

**2.2.2.2 Folding structures.** It has been established that salt responses like a pseudo-fluid to the stress at geological time scale that results in ductile or creep deformation rather than in brittle behavior (Anderson and Brown, 1992; Brady and Brown, 2007; Warren, 1989). Extensive folding of variable amplitudes and wavelengths is characteristic of evaporite formation of the Upper Kama deposit (Kudryashov, 2001).

The major folding structures of the evaporites in the Upper Kama deposit have a submeridianal orientation, perpendicular to principal lateral E-W tectonic stress caused by the Uralian Orogen (Kudryashov, 2001) that is confirmed by westward inclined folds axis. The deformations are stronger at a west flank of deposit resulting in number of submeridianal brachianticlines extending from north to south end of deposit (Figure 2.11b). The presence of submeridianal brachianticlines, presumably, resulting from the principal East-West stress, suggests that the salts were extensively deformed after deposition.

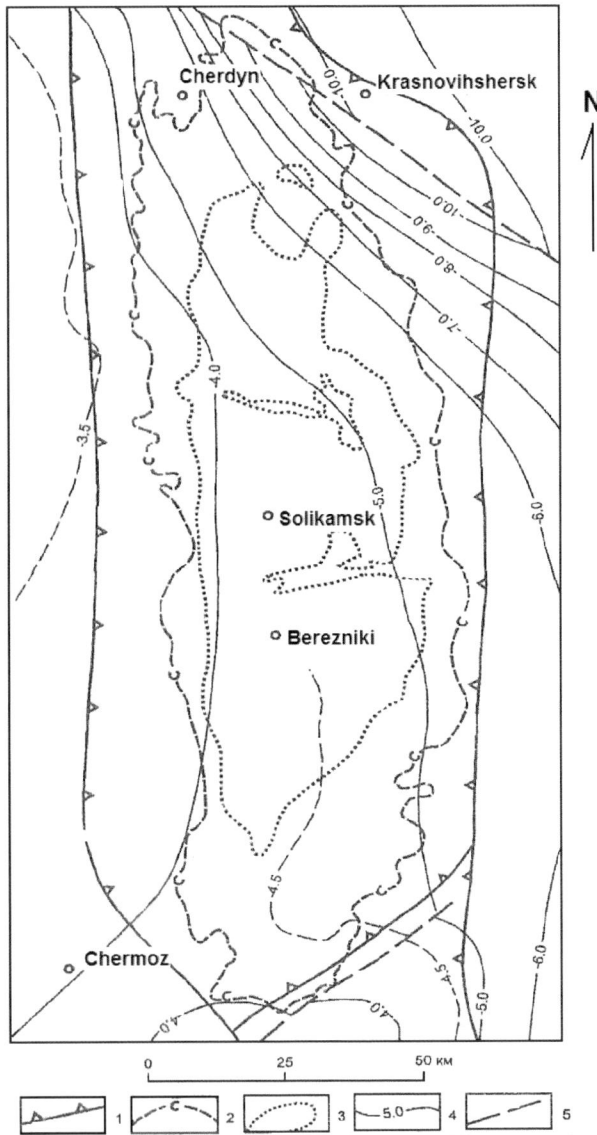

Figure 2.10. Map of structure of crystalline basement, Solikamsk depression (modified from Kudryashov, 2001). Explanation: 1 – boundaries of the Solikamsk depression; 2 – boundary of the Upper Kama salt deposit; 3 – boundary of potash deposit; 4 – contour lines of basement depth; 5 – faults.

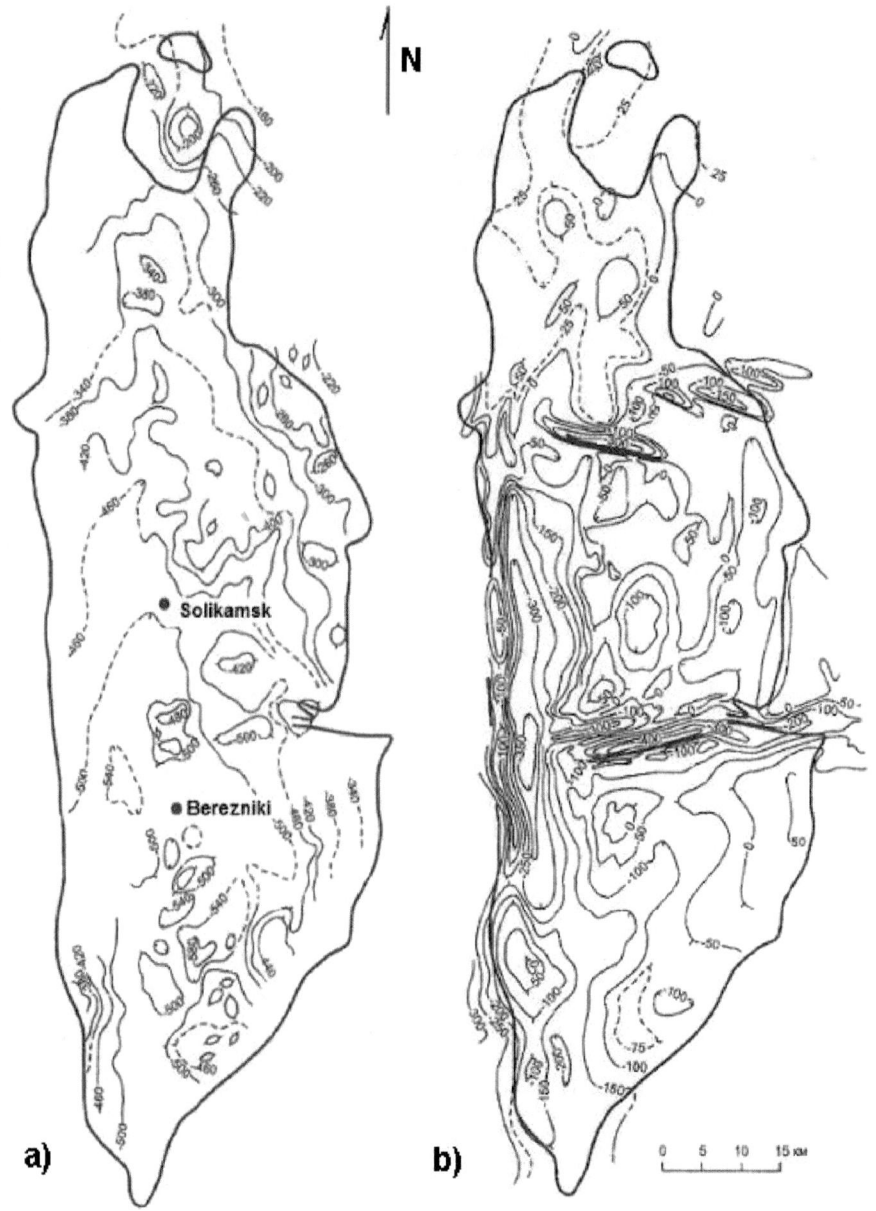

Figure 2.11. Structural maps of bottom of evaporite formation (a) and top of the

overlying rock salt (b) (modified from Kudryashov, 2001). The latitudinal deep structures

present the areas of salt dissolution at regional fault zones.

Locally, a superposition of a major tectonic stress and stresses from the faults and numerous uplift structures represented mainly by reef buildups leads to reorientation of deformation structures or overlapping the structures of different trend (Konstantinova et al., 2001). Comparison of the structural maps of the bottom surface of evaporite formation and a top of Overlying Rock Salt (see Figures 2.11a and 2.11b) shows that some salt structures follow the underlying beds relief but many features are the result of salt deformation, halokinetic and dissolution processes.

Because of salt plasticity the bending deformations with movement of beds along planes of stratification and flowage structures are often observed. The flow folding structures, often overturned, occur mostly in the sylvinite and carnallite layers. The typical folding structures in the interval of sylvinite layers are presented in Figures 2.12a, 2.12b, and 2.13.

**2.2.2.3 Brittle deformation of potash salts.** Brittle deformation is not common for salt rocks and occurs locally in the evaporite formation of the Upper Kama deposit where tectonic, gravitational or mine induced stress exceeded the strength of the rock mass (Brady and Brown, 2007; Kudryashov et al., 2004; Lajtai et al., 1994).

Geological and geophysical studies revealed the systems of regional faults of longitudinal, latitudinal, northwest and northeast orientation (Kudryashov et al., 2004). Schematic map of fault tectonics of potash deposit area is depicted in Figure 2.14.

The large solution structures, such as Durinskaya and Borovitskaya depressions (Figure 2.14), and collapse zone at Berezniki mine 3, are related to the faults in salt formation and overburden sediments (Andrejchuk, 2002; Kudryashov et al., 2004).

Figure 2.12. Fold structures in sylvinite layers (a) and in clay layer (b) (from
Chaykovskiy et al., 2009).

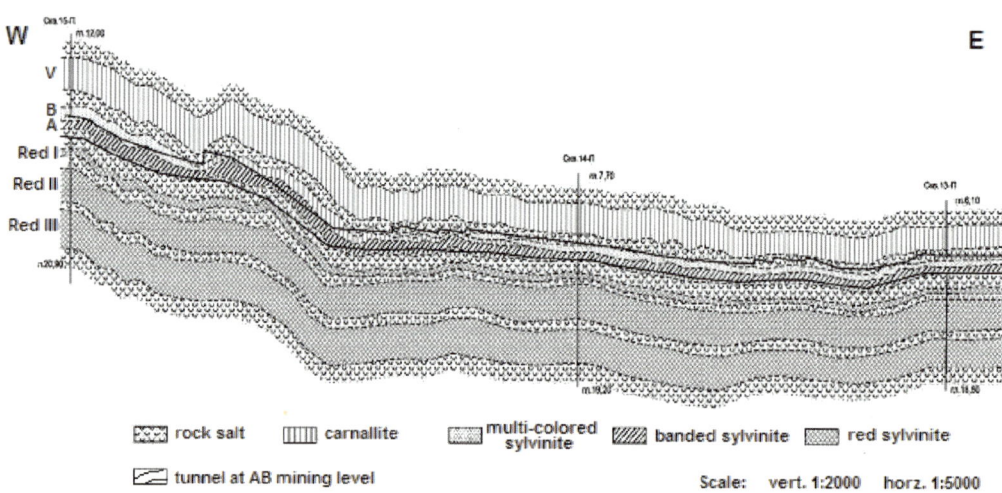

Figure 2.13. W-E cross-section of sylvinite zone at the Panel 18 of Berezniki potash mine
2. Note the height of the mine tunnel is about 3 m (modified from Belkin, 2010).

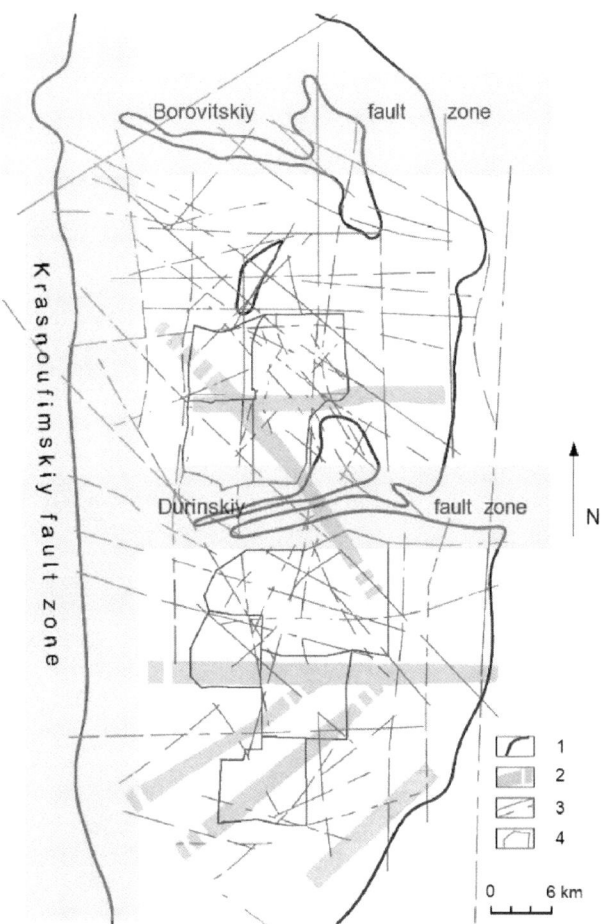

Figure 2.14. Schematic map of fault structures at the area of the Upper Kama potash deposit (modified from Kudryashov et al., 2004). Explanation: 1 – boundaries of potash deposit; 2 – regional and local fault zones relatively to the area of influence; 3 – separate faults; 4 – mine fields.

Three regional thrust faults were identified within the evaporite formation (Figure 2.15). However, the existence of some thrusts is still debated because of difficult identification affected by intensive healing process in salt rock (Kudryashov et al., 2004; Jeremic, 1995).

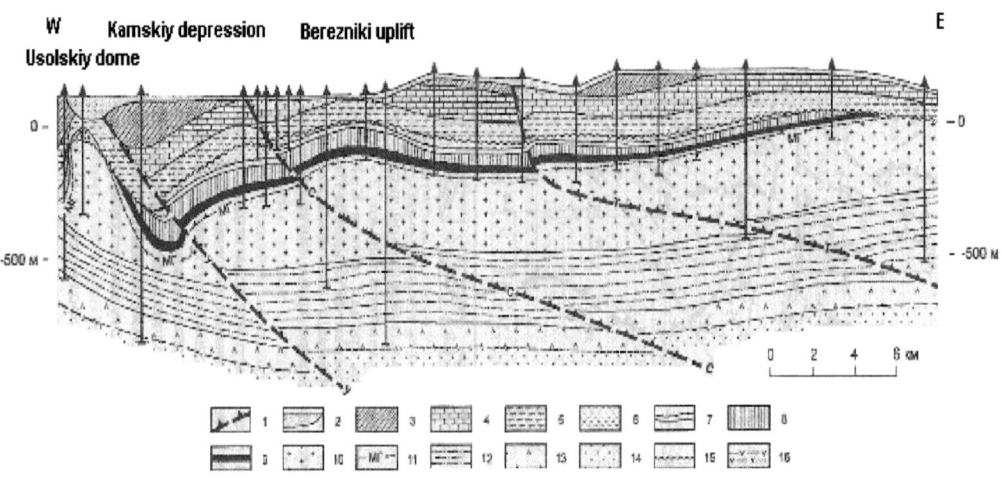

Figure 2.15. Latitudinal geological cross-section showing location of the thrust faults established at the Upper Kama potash deposit (modified from Kudryashov, 2001). Explanation: 1 – thrust faults; 2 – Cenozoic clay; 3 – multi-colored complex; 4 – terrigeneous-carbonate complex; 5 – upper member of salt-marl complex; 6 – transition sequence; 7 – overlying rock salt; 8 – sylvinite-carnallite zone; 9 – sylvinite zone including B layer; 10 – underlying rock salt; 11 – Marker Clay; 12 – predominantly aleurolite of Irenian suite; 13 – anhydrites of Karnaukhovskaya suite; 14 – Artinskian terrigeneous sediments; 15 – washout surface on a top of salt formation; 16 – gypsum-clay caprock.

A number of discrete natural fractures and fracture systems of various scales have been observed during mine operations, especially, in a central part of deposit. Most fractures within the salts are healed and are visually observed only in the clay-anhydrite layers. Open fractures are generally encountered within the sylvinite-carnallite zone. They develop in meridianal and NW-SE directions. Locally, a centimeter-scale vertical displacement is observed. Typical fractures observed in the potash mines of the Upper Kama deposit are shown in (Figure 2.16).

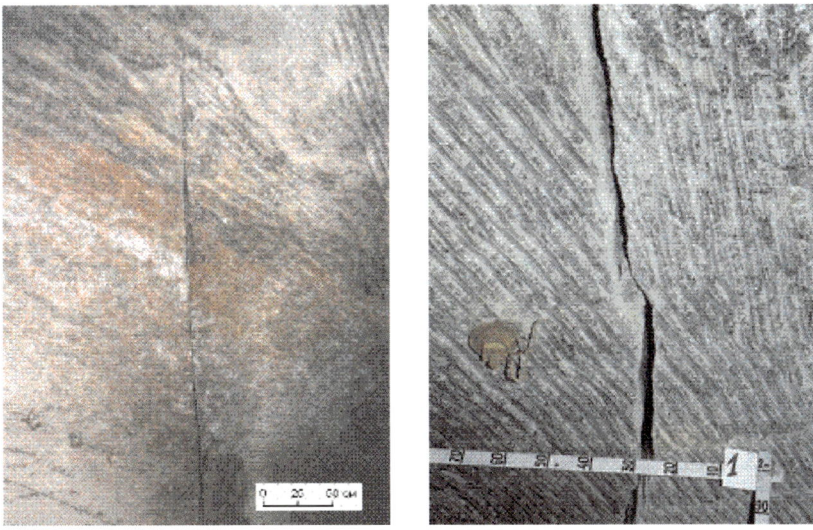

Figure 2.16. Typical fractures encountered in salts of the Upper Kama potash mines (from Kudryashov et al., 2004).

Predominantly, fractures are S-shaped and do not extend beyond the individual layer. The most intense deformation is noted in the carnallite zone, where very steep folding, tectonic breccia and blocks of interbed rock salt are observed during the mining of carnallite layers and (Kudryashov et al., 2004).

The largest set of fractures discovered in the Upper Kama potash deposit was exposed at the eastern edge of mining field of the Solikamsk mine 3 (Figure 2.17).

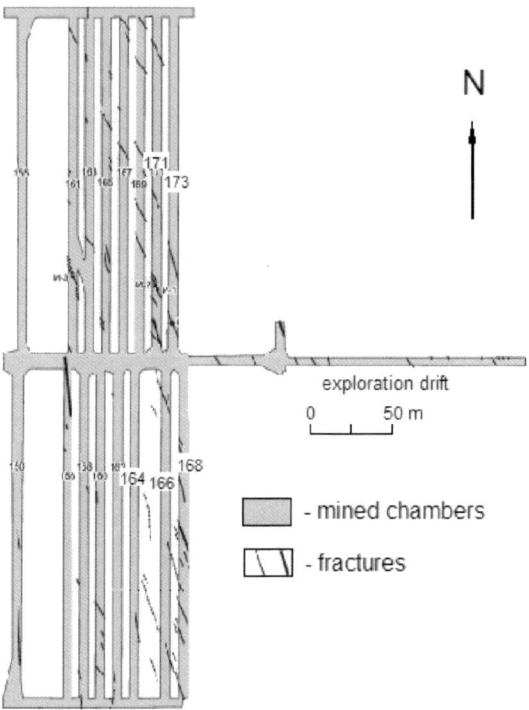

Figure 2.17. Fracture zone at the Panel 4 of Solikamsk mine 3 (modified from Kudryasov et al., 2004).

Numerous oblique subvertical tension fractures were encountered in the excavations of mining level AB. The fractures were observed at the total area 400 to 100 m (Figure 2.17). Opening of the fractures striking NW 335°- 355° ranges from less than one millimeter to more than one centimeter. Only a few fractures were observed 8-9 m below level AB at the mining level Red II. This suggests that the vertical extent of fracture zone is about 10 m. An example of centimeter scale fracture is shown in Figure 2.18.

Figure 2.18. Cascade fractures of centimeter scale opening exposed on the wall of working at mining level AB, Panel 4, Solikamsk mine 3.

Upward exploration boreholes and visual observation of the roof showed that the fractures die out within overlying carnallite layer or interlayer clay seams (Figure 2.19).

Figure 2.19. Exploration borehole driven in the roof intersects a fracture exposed at the opening of mining level AB, Panel 4, Solikamsk mine 3. It is seen that the fracture becomes thinner in upward direction.

All the fractures intersect the NE-SW trending folds and, probably, are the result of local stress redistribution caused by NW-SE fault structure located eastward from the site (Kudryashov et al., 2004).

## 2.3. MINING SETTINGS AND PROBLEMS

**2.3.1. Mining Operation.** Currently five underground mines produce potash ore at the Upper Kama deposit (see Figure 2.2). Also carnallite is mined at the Solikamsk mine 1 for extraction of magnesium, as well as some amount of halite is mined for table and technical salt.

All of the mines use a multiple-level room-and-pillar method of mining, and the rooms are 200 m length, 3 m to 15 m in width and 3 m to 10 m high. Combination of stiff and soft (yield) pillar mining, with purposefully subsiding pillars is often practiced (Nesterov, 1995). Pillars between the rooms are of 3-18 m in width. Depending on the number of extracted sylvinite layers, the mining levels are separated by 5-15 m of interbed rock salt.

The sylvinite ore is mined using cutting machines. Blasting is used for mining the carnallite layers because of their high gas content.

In the Solikamsk mine 3, potash ore is extracted from a depth of about 300 m. The two most sylvite-rich beds, AB and Red II, were mined at the author's research survey site (block 2 of south-east panel 4). The ventilation and conveyor tunnels are excavated under the potash layers in the underlying rock salt (Figure 2.20).

Water protective beds about 130 m thick isolate the mine openings from the overlying aquifers. Usually, there is a salt-saturated brine layer, referred to as brine horizon, lying on top of uppermost rock salt layer protecting it from leaching. A number of salt-unsaturated aquifers with water pressures of up to 3.5 MPa are located in fractured or clastic beds of the overlying salt-marl and terrigeneous-carbonate complexes (Garrett, 1995).

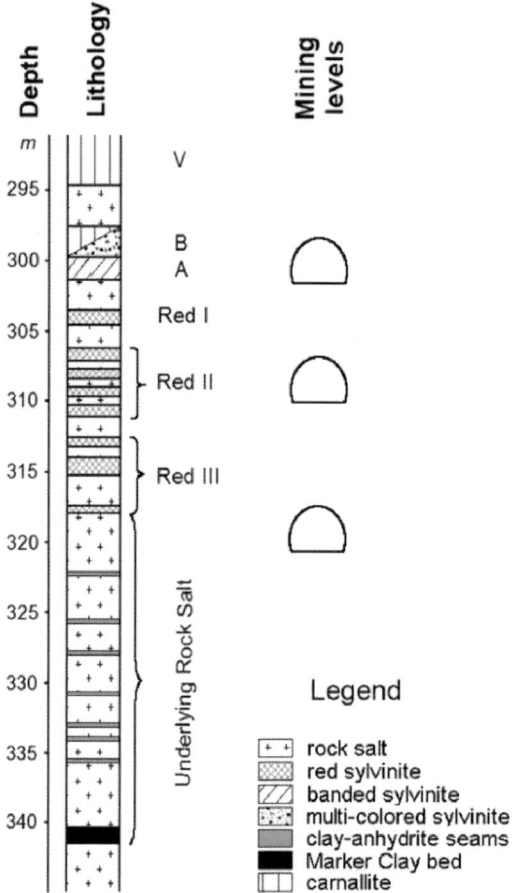

Figure 2.20. Generalized scheme of mining settings at Panel 4, Solikamsk mine 3.

Openings at AB and Red II mining levels are the production excavations, and lower

excavation, in the underlying rock salt, is used for ore and personnel transportation, and

ventilation purposes.

**2.3.2. Mining Problems.** Salt is a highly soluble material. The prevention of the salt-unsaturated water inflow into a salt mine is therefore a priority. The sudden failure of supporting pillars or the reopening of natural fractures during mining operation can create pathways for groundwater inflow into the mine. Salt is about 7500 times more soluble than limestone (Martinez et al., 1998). This enables salt-unsaturated water to form pathways through the otherwise impermeable protective salt beds in a very short time.

Undermined areas of the Upper Kama (Verkhnekamskoye) potash deposit within the cities of Berezniki and Solikamsk (Perm region, Russia) represent a potential hazard because water influx and mine failure is a possibility, however remote. Large numbers of people, chemical and power plants, roadways, railroads and gas pipelines are located above the area of the mine.

In 1986 the Berezniki potash mine 3 was lost after the sudden failure of supporting pillars induced brittle deformation of the overlying impermeable water protective cover with subsequent catastrophic flooding (Andrejchuk, 2002; Whyatt and Varley, 2008). A large collapse structure was created on the surface (Figure 2.21).

In 1995, the sudden failure of pillars and resultant roof collapse in the area 600 to 700 m occurred at the Solikamsk mine 2. The failure of the supporting pillars was accompanied by subsidence of the surface ranging from 3 to 4.5 m (Konstantinova, 1999). Subsidence map of the area above the collapse zone in Solikamsk mine 2 is shown in Figure 2.22. A roof fall seismic event magnitude of 4.7 was recorded (Malovichko et al., 2001, 2005). Fortunately, no water influx to the mine was observed and the hazardous area was later backfilled.

Significant financial losses were incurred by the JSC Uralkali and state due to replacement/repair costs of roadways and other infrastructure after the catastrophic collapse at the Berezniki potash mine 1 in 2006 (Figure 2.23).

Figure 2.21. Collapse structure 210 to 140 m in size was created above the water pathway at Berezniki potash mine 3. Long axis of the structure is oriented to direction of the strike of fault in overlying strata.

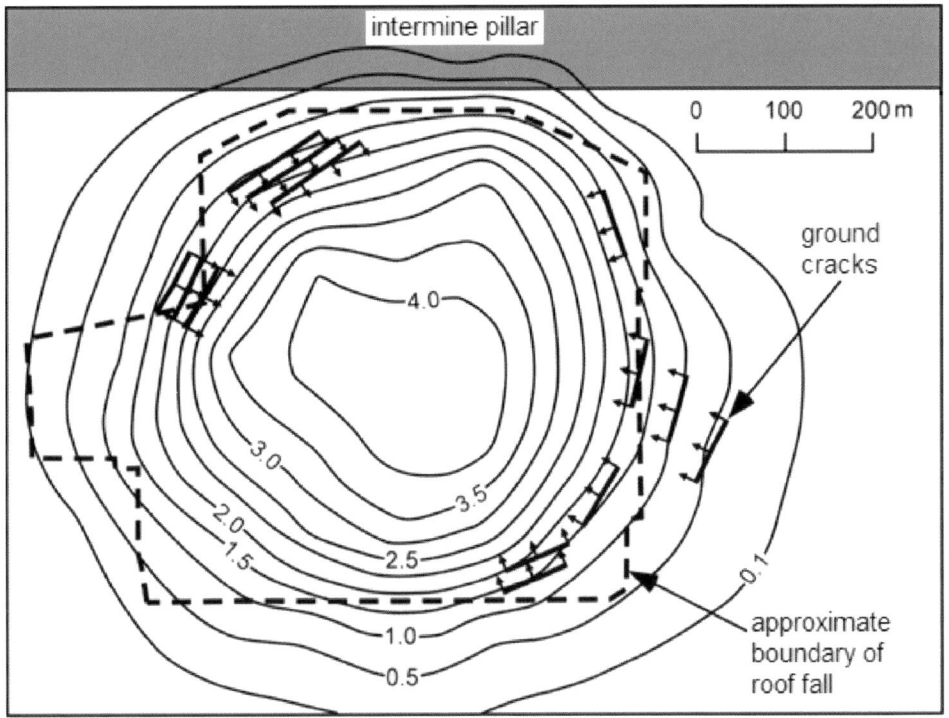

Figure 2.22. Subsidence map at the area above collapse zone in Solikamsk potash mine 2 (modified from Malovichko et al., 2001).

Figure 2.23. Collapse structure occurred at Berezniki potash mine 1 in 2007 after the water influx into the mine initiated in 2006. The size of structure achieved 446 to 335 m in 2009. Sinkhole cut the railroad Perm – Berezniki that can be seen on the front of collapse structure (from Geography, 2009).

There are tens of flooded potash, salt and trona mines around the world (Whyatt and Varley, 2008; Zipf and Swanson, 1999; Neal and Myers, 1995; Martinez et al., 1998; Prugger, 1980; Prugger and Prugger, 1991). All of the cases are related to inflow of groundwater into the mine after conduits were developed in impermeable barrier as a result of sudden roof fall or the reopening pre-exist natural fractures and geological anomalies.

Stability of underground mine openings depends on factors such as the manner of mining, the geomechanical properties of the supporting pillars and the roof beds, the conditions of water protective cover, and subsidence preventive measures employed (such as the backfilling of the mine workings with potash ore post-processing wastes).

According to regulations for the operations of the Upper Kama potash mines, open fractures and high-amplitude overturned folds are assumed to be prerequisite for failure. Such structures/features, where identified, must be thoroughly investigated to minimize the risk of roof rock failure and catastrophic flooding.

Seismic, electrical, gravity and seismological methods are used to monitor the water protective strata at the potash mines of the Upper Kama deposit (Malovichko et al., 2005; Glebov, 2006). Anomalous objects in vicinity of mine workings and the condition of the mine pillars are routinely monitored by using underground seismic, acoustic and electrical methods. Geomechanical control is provided by measurements of stress and strain parameters in the mines and by laboratory mechanical testing of samples. Numerical simulation is used to predict the rock mass behavior in certain mine environments.

After two mines were flooded and lost, the mining companies searched for new mining locations, and encountered a new challenge. Petroleum companies were exploring for hydrocarbons in sub-salt Devonian and Visean strata. This posed a new problem for the mining companies because the oil boreholes/wells could trigger flooding (initiated by leakage through boreholes, corroded casing) and reservoir exploitation causes the stress redistribution in the surrounding rock mass. The mining and oil companies jointly developed measures such as horizontal drilling, preventive pillars around the wells, usage of special cementation methods, and permanent instrumental control about the parameters of surrounding rock mass to protect the interests of both groups of companies.

# 3. THEORY OF GEORADAR METHOD

## 3.1. FUNDAMENTALS OF ELECTROMAGNETIC WAVES

**3.1.1. Maxwell's Equations.** The theory of electromagnetic fields is based on the solution of the Maxwell's equations. These equations state the basic principles of radio waves propagation in a medium. Maxwell's equations describe the interaction between electric and magnetic fields and the relationship between the charge and current density (Griffiths, 1999). They can be written for source-free homogeneous medium in differential form as:

$$\nabla \times \mathbf{E} = -\frac{\partial \mathbf{B}}{\partial t} \tag{1}$$

$$\nabla \times \mathbf{H} = \mathbf{J} + \frac{\partial \mathbf{D}}{\partial t} \tag{2}$$

$$\nabla \bullet \mathbf{B} = 0 \tag{3}$$

$$\nabla \bullet \mathbf{D} = \rho \tag{4}$$

where

$\mathbf{E}$ – electric field intensity in $V/m$

$\mathbf{B}$ – magnetic induction in $Wb/m^2$ or $Tesla$

$\mathbf{H}$ – magnetic field intensity in $A/m$

$\mathbf{J}$ – electric current density in $A/m^2$

$\mathbf{D}$ – electric induction in $C/m^2$

$\rho$ – electric charge density in $C/m^3$.

A set of two vector and two scalar Maxwell's equations for electrostatic field integrates: Faraday's law for magnetism (1), Ampere's law (2), Gauss's law for magnetism (3), and Gauss's law for electric field (4).

**3.1.2. Constitutive Relations.** To apply Maxwell's equations to the description of electromagnetic fields in the real materials, information about electric and magnetic material properties is required. Four constitutive equations relate the material properties to the response from electromagnetic field applied to these materials:

$$\mathbf{J} = \sigma\mathbf{E} \tag{5}$$

$$\mathbf{D} = \varepsilon\mathbf{E} \tag{6}$$

$$\mathbf{B} = \mu\mathbf{H} \tag{7}$$

$$\mathbf{M} = \chi\mathbf{H} \tag{8}$$

where

$\sigma$ - electric conductivity in $S/m$

$\varepsilon$ - dielectric permittivity in $F/m$

$\mu$ - magnetic permeability in $H/m$

$\mathbf{M}$ - magnetization

$\chi$ - magnetic susceptibility.

Four quantities, namely $\sigma$, $\varepsilon$, $\mu$, and $\chi$, exclusively describe the properties of a material. The three first relations are the most important for GPR (Annan, 2001). In the general case, $\sigma$, $\varepsilon$ and $\mu$ are not constants. They depend on the strength, direction and frequency of the fields, and the spatial inhomogeneity of the material. In the case of the GPR method, these values are assumed independent of the parameters of existing fields (Annan, 2001). Understanding of the behavior of $\sigma$ and $\varepsilon$ under influence of electromagnetic fields is of most importance for understanding the results of the GPR survey.

For the practical purposes of mathematical calculations and to satisfy SI unit system, the dielectric permittivity $\varepsilon$ and magnetic permeability $\mu$ usually are defined as a product of the dielectric permittivity and magnetic permeability in a vacuum ($\varepsilon_0$, $\mu_0$) and the relative permittivity or *dielectric constant* ($\varepsilon_r$) and relative permeability ($\mu_r$) respectively:

$$\varepsilon = \varepsilon_0 \varepsilon_r \tag{9}$$

$$\mu = \mu_0 \mu_r \tag{10}$$

where the electric constant $\varepsilon_0$ and magnetic constant $\mu_0$ are:

$$\varepsilon_0 = 8.854... \times 10^{-12} \left( F/m \right)$$

$$\mu_0 = 4\pi \times 10^{-7} \left( H/m \right).$$

**3.1.3. Electromagnetic Waves.** Two equations, which constitute the Faraday's Law (1) and Ampere's Law (2), describe the time-variant interaction of coupled electric and magnetic fields. Applying to these equations the constitutive relations and relations from vector field theory yields the explicit expressions for electromagnetic wave propagation.

If one takes the curl of (1), and use expressions (2), (5), and (6), we have for the electric field:

$$\nabla \times \nabla \times \mathbf{E} = -\mu \frac{\partial}{\partial t}(\nabla \times \mathbf{H}) = -\mu \frac{\partial}{\partial t}\left(\sigma \mathbf{E} + \varepsilon \frac{\partial \mathbf{E}}{\partial t}\right) = -\mu\sigma \frac{\partial \mathbf{E}}{\partial t} - \mu\varepsilon \frac{\partial^2 \mathbf{E}}{\partial t^2}. \qquad (11)$$

Taking the curl of (2) and inserting (1), (5), and (6) give the similar expression for the magnetic field:

$$\nabla \times \nabla \times \mathbf{H} = \sigma(\nabla \times \mathbf{E}) + \varepsilon \frac{\partial}{\partial t}(\nabla \times \mathbf{E})$$
$$= \sigma\left(-\mu \frac{\partial \mathbf{H}}{\partial t}\right) + \varepsilon \frac{\partial}{\partial t}\left(-\mu \frac{\partial \mathbf{H}}{\partial t}\right). \qquad (12)$$
$$= -\mu\sigma \frac{\partial \mathbf{H}}{\partial t} - \mu\varepsilon \frac{\partial^2 \mathbf{H}}{\partial t^3}$$

Applying the vector identity relationship $\nabla \times \nabla \times \mathbf{A} = \nabla(\nabla \bullet \mathbf{A}) - \nabla^2 \mathbf{A}$ for any given vector field $\mathbf{A}$ to the obtained relations, yields the instantaneous vector wave equations (Helmholtz equations):

$$\nabla^2 \mathbf{E} = \mu\sigma \frac{\partial \mathbf{E}}{\partial t} + \mu\varepsilon \frac{\partial^2 \mathbf{E}}{\partial t^2} \qquad (13)$$

$$\nabla^2 \mathbf{H} = \mu\sigma \frac{\partial \mathbf{H}}{\partial t} + \mu\varepsilon \frac{\partial^2 \mathbf{H}}{\partial t^2}. \qquad (14)$$

Finally, we have two equations decoupled; only one component of electromagnetic field is present in one equation. Introducing the concept of *phasors* (Hayt, 1989) we may transform the instantaneous (time-domain) vector $\mathbf{A}$ into phasor (frequency-domain) vector $\mathbf{A_s}$ as:

$$\mathbf{A} \Leftrightarrow \mathbf{A_s}$$

$$\frac{\partial \mathbf{A}}{\partial t} \Leftrightarrow j\omega\mathbf{A_s}$$

$$\frac{\partial^2 \mathbf{A}}{\partial t^2} \Leftrightarrow (j\omega)^2 \mathbf{A_s}$$

where $j = \sqrt{-1}$ and the quantities become complex and are written as .

Assuming the harmonic variation of electromagnetic field at circular frequency $\omega$, the Helmholtz equations 11 and 12 can be written in following form:

$$\nabla^2 \mathbf{E_s} = j\omega\mu(\sigma + j\omega\varepsilon)\mathbf{E_s} \qquad (15)$$

$$\nabla^2 \mathbf{H_s} = j\omega\mu(\sigma + j\omega\varepsilon)\mathbf{H_s}, \qquad (16)$$

where $\mathbf{E_S}$ and $\mathbf{H_S}$ are the electric field and magnetic field phasors.

If we assume $j\omega\mu(\sigma + j\omega\varepsilon) = \gamma^2$ the Helmholtz equations will reduce to the form of the phasor vector wave equations (Baker and Jol, 2007):

$$\nabla^2\mathbf{E_s} - \gamma^2\mathbf{E_s} = 0 \tag{17}$$

$$\nabla^2\mathbf{H_s} - \gamma^2\mathbf{H_s} = 0. \tag{18}$$

The complex constant $\gamma$ is termed the *propagation constant* $(m^{-1})$ and can be presented in more convenient form for manipulation as:

$$\gamma = \sqrt{j\omega\mu(\sigma + j\omega\varepsilon)} = \alpha + j\beta, \tag{19}$$

where the real part $\alpha$ is defined as the *attenuation constant* (*Nepers/m*) and the imaginary part $\beta$ is defined as the *phase constant* (*Rad/m*). Equations for attenuation and phase constants can be determined from (15):

$$\alpha = \omega\sqrt{\frac{\mu\varepsilon}{2}\left[\sqrt{1+\left(\frac{\sigma}{\omega\varepsilon}\right)^2} - 1\right]} \tag{20}$$

$$\beta = \omega\sqrt{\frac{\mu\varepsilon}{2}\left[\sqrt{1+\left(\frac{\sigma}{\omega\varepsilon}\right)^2} + 1\right]}. \tag{21}$$

**3.1.3.1 Concept of complex permittivity.** Permittivity is very important property of material and defines the applicability of the GPR method in a given environment. Permittivity quantitatively shows how a material is capable of bieng polarized and accumulating energy in response to an external electric field (Olhoeft, 1998).

As stated by Ampere's Law (2), a current generated in the media in response to an external magnetic field consists of conduction and displacement currents. This may be written in the form:

$$\nabla \times \mathbf{H_s} = \mathbf{J}_C + \mathbf{J}_D = \sigma \mathbf{E_s} + j\omega\varepsilon\mathbf{E_s} = (\sigma + j\omega\varepsilon)\mathbf{E_s}$$
$$= j\omega\varepsilon\left(1 + \frac{\sigma}{j\omega\varepsilon}\right)\mathbf{E_s} = j\omega\varepsilon\left(1 - j\frac{\sigma}{\omega\varepsilon}\right)\mathbf{E_s} \quad , \tag{22}$$

where

$\mathbf{J}_C$ – conduction current density in $A/m^2$

$\mathbf{J}_D$ – displacement current density in $A/m^2$.

Quantity $\varepsilon_c = 1 - j\sigma/\omega\varepsilon$ from above equation is defined as a complex permittivity:

$$\varepsilon_c = \varepsilon' - j\varepsilon'' = \varepsilon_c e^{-j\theta}, \tag{23}$$

where $\theta$ (referred to as loss angle) characterizes a ratio of stored to dissipated energy (see Figure 3.1). The most common term in the GPR context is the loss tangent that is a relationship between conduction and displacement current (Ward and Hohmann, 1988):

$$\frac{|\mathbf{J}_C|}{|\mathbf{J}_D|} = \frac{|\sigma\mathbf{E}_C|}{|j\omega\varepsilon\mathbf{E}_C|} = \frac{\sigma}{\omega\varepsilon} = \frac{\varepsilon''}{\varepsilon'} = \tan\theta \,. \tag{24}$$

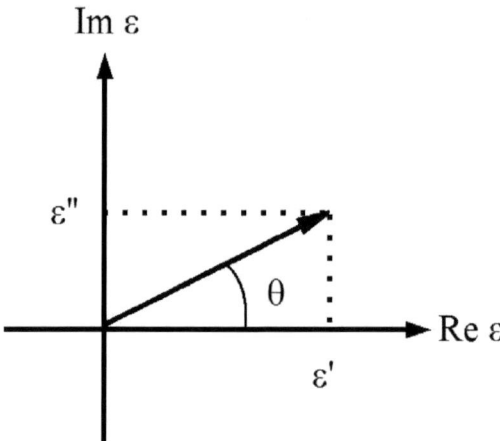

Figure 3.1. Vector diagram of real and imaginary parts of complex permittivity, and loss tangent θ.

The displacement currents depend on frequency and this factor is important for estimating GPR effectiveness in certain environments. A simplified plot of conduction, displacement and total current $\mathbf{J} = \mathbf{J}_C + \mathbf{J}_D$ versus frequency is shown in Figure 3.2. One can see that there is a frequency range where the displacement currents exceed the conduction currents and there is a transition frequency $\omega_t$ where the currents are equal. This frequency defines the low-loss frequency range for GPR (Annan, 2001).

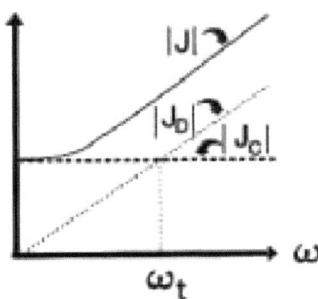

Figure 3.2. Chart of total, conduction and displacement currents versus frequency (from Annan, 2001).

The transition frequency for dielectric material may be defined as

$$\omega_t = \frac{\sigma}{\varepsilon_0 \varepsilon_r} \quad \text{and} \quad f_t = \frac{\sigma}{2\pi\varepsilon_0 \varepsilon_r}, \tag{25}$$

where

$\omega_t$ – circular transition frequency (Rad/m)

$f_t$ – transition frequency (Hz).

It should be noted that the transition frequency depends on both conductivity and relative permittivity.

**3.1.3.2 Plane waves.** It is convenient for practical purposes to use the concept of a plane wave. In the case of a plane wave, **E** and **H** lie in a plane perpendicular to the direction of propagation and are orthogonal to each other. The fields **E** and **H** vary only in the direction of propagation.

In Cartesian coordinate, for harmonic plane wave propagating into the direction of the *y*-axis, the Helmholtz equations reduce to the linear differential equations:

$$\frac{d^2 E_{ZS}}{dy^2} - \gamma^2 E_{ZS} = 0 \tag{26}$$

$$\frac{d^2 H_{XS}}{dy^2} - \gamma^2 H_{XS} = 0, \tag{27}$$

where $E_{ZS}$ and $H_{XS}$ are functions of *y* only.

The direction of electric field defines the *polarization* of plane electromagnetic wave (Figure 3.3). The electromagnetic waves having no components in the direction of propagation are termed *transverse electromagnetic waves*.

To find the solution of wave equation for arbitrary directed plane wave we define a *vector wave-number* or *propagation vector* **k** whose direction coincides with direction of wave propagation and whose magnitude in case of lossless media is $\omega\sqrt{\mu\varepsilon}$ .

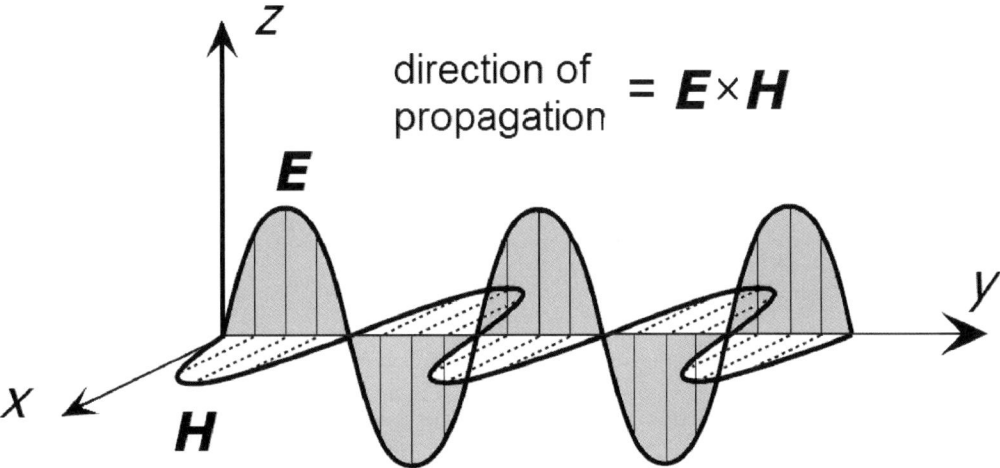

Figure 3.3. Propagation of electromagnetic wave in the direction of *y*-axis in the "lossless" medium. Z-polarized plane wave has only a z-component of electric field and x-component of magnetic field, which are the function only of *y*.

Introducing position vector **r** we can write:

$$\mathbf{k} = \mathbf{k}_x + \mathbf{k}_y + \mathbf{k}_z$$
$$\mathbf{r} = \mathbf{r}_x + \mathbf{r}_y + \mathbf{r}_z$$

$$\mathbf{E}_S = E_{S0}e^{-j(\mathbf{k}\cdot\mathbf{r})}\mathbf{a}_E = E_{S0}e^{-j(k_x x + k_y y + k_z z)}\mathbf{a}_E \tag{28}$$

$$\mathbf{H}_S = H_{S0}e^{-j(\mathbf{k}\cdot\mathbf{r})}\mathbf{a}_H = H_{S0}e^{-j(k_x x + k_y y + k_z z)}\mathbf{a}_H, \tag{29}$$

where $\mathbf{a}_E$ and $\mathbf{a}_H$ are the unit vectors of the electric and magnetic fields, respectively.

## 3.2. PROPAGATION OF ELECTROMAGNETIC WAVES IN THE MEDIUM

To define the behavior of an electromagnetic wave in a media, the solution of Helmholtz equation should be found. In the case of an electromagnetic wave traveling in the $y$-axis direction, the solution is given as:

$$E_{zs}(y,\omega)= E_0 e^{-\gamma y} = E_0 e^{-\alpha y} e^{-j\beta y} \tag{30}$$

$$H_{zs}(y,\omega)= H_0 e^{-\gamma y} = H_0 e^{-\alpha y} e^{-j\beta y}, \tag{31}$$

where $E_0$ and $H_0$ are the scalar amplitudes of electric and magnetic fields. The analogous time-domain solution can be found in the form:

$$E_z(y,t)=E_0 e^{-\alpha y} \cos(\omega t - \beta y) \tag{32}$$

$$H_z(y,t)= H_0 e^{-\alpha y} \cos(\omega t - \beta y) \tag{33}$$

**3.2.1. Properties of Electromagnetic Waves.** Phase velocity, attenuation and electromagnetic impedance are the most important properties of electromagnetic waves in terms of GPR applications (Annan, 2001). At low frequencies, all of the properties are frequency dependent and the electromagnetic field has a diffusive character. The radio signals are highly dispersed and require interpretation in term of electromagnetic induction. At high frequencies (> 10-15 MHz), the electromagnetic waves propagate as waves.

Solving equations (32) and (33) for the variable $y$, one can get the position of constant phase point in the form:

$$y = \frac{1}{\beta}\left(\omega t - const\text{ant}\right).$$ (34)

Differentiating (34) with respect to time yields the velocity at which the constant phase point moves along $y$-axis:

$$v = \frac{\omega}{\beta}$$ (35)

where

$v$ – phase velocity in $m/s$.

The other important relations for electromagnetic wave propagation characterization are:

$$\omega = 2\pi f = \frac{2\pi}{T}$$ (36)

$$\lambda = vT = \frac{v}{f} = \frac{\omega/\beta}{f} = \frac{2\pi}{\beta}$$ (37)

where

$f$ – frequency in $Hz$

$T$ – period in $s$

$\lambda$ – wavelength in $m$.

The rate of attenuation can be characterized by the *skin-depth* $\delta$ that is a distance at which a plane wave has been attenuated by a factor $e^{-1}$:

$$\delta = \frac{1}{\alpha} = \frac{1}{\sqrt{\pi f \mu \varepsilon}}.\qquad(38)$$

The amplitudes of electric and magnetic fields can be related to each other because the fields are coupled according to Maxwell's equations. *Electromagnetic impedance Z* is defined as a ratio of complex amplitudes of the electric field and magnetic field phasors:

$$Z = \frac{E_s}{H_s} = \frac{j\omega\mu}{\gamma} = \sqrt{\frac{j\omega\mu}{\sigma + j\omega\varepsilon}}.\qquad(39)$$

Electromagnetic impedance is a complex number. Its magnitude is:

$$|Z| = \frac{\sqrt{\mu/\varepsilon}}{\left[1+\left(\dfrac{\sigma}{\omega\varepsilon}\right)^2\right]^{-\frac{1}{4}}}.\qquad(40)$$

**3.2.1.1 Solution for lossless media.** In the lossless media (perfect dielectrics), EM waves propagate without attenuation ($\alpha = \sigma = 0$). Using (15), (16) and (17) one may find that

$$\alpha = 0 \quad \text{and} \quad \beta = \omega\sqrt{\mu\varepsilon}\,. \tag{41}$$

Consequently, the velocity of electromagnetic wave propagation and impedance in lossless media can be found in the following form:

$$v = \frac{\omega}{\beta} = \frac{1}{\sqrt{\mu\varepsilon}} \tag{42}$$

$$Z = \sqrt{\frac{\mu}{\varepsilon}}\,. \tag{43}$$

Using expressions (9), (10) and (42), the velocity $c$ of wave propagation in vacuum (free space) may be defined as

$$c = \frac{1}{\sqrt{\mu_0\varepsilon_0}} \approx 2.998\cdot10^8 \ m/s\,. \tag{44}$$

**3.2.1.2 Solution for lossy media.** In case of conductive media ($\sigma > 0$), the complex propagation constant can be defined using (19) as (Ward and Hohmann, 1988)

$$\gamma = j\omega\sqrt{\mu\varepsilon}\sqrt{1 - j\tan\theta} = \alpha + j\beta\,. \tag{45}$$

The attenuation and phase constants may be written in the following form:

$$\alpha = \omega \sqrt{\frac{\mu\varepsilon}{2}\left[\sqrt{1+\left(\frac{\sigma}{\mu\varepsilon}\right)^2}-1\right]} = \omega\sqrt{\frac{\mu\varepsilon}{2}}\left[\sqrt{1+\tan^2\theta}-1\right] \qquad (46)$$

$$\beta = \omega \sqrt{\frac{\mu\varepsilon}{2}\left[\sqrt{1+\left(\frac{\sigma}{\mu\varepsilon}\right)^2}+1\right]} = \omega\sqrt{\frac{\mu\varepsilon}{2}}\left[\sqrt{1+\tan^2\theta}+1\right]. \qquad (47)$$

Velocity of electromagnetic wave propagation is defined as:

$$v = \frac{\omega}{\beta} = \frac{1}{\sqrt{\mu\varepsilon}} = \frac{c}{\sqrt{\mu_r\varepsilon_r}} \qquad (48)$$

and the electromagnetic impedance becomes complex number:

$$Z = \sqrt{\frac{j\omega\mu}{\sigma+j\omega\varepsilon}}. \qquad (49)$$

The electric and magnetic fields in lossy medium are out of phase alternatively to the case of lossless medium. Because the attenuation, predominantly related to conductivity, is not zero and the propagation constant is frequency dependent, the amplitude of the wavelet decays exponentially in the direction of travel and the shape of signal changes.

The case of small losses is of most interest in practical GPR applications because all real materials are lossy dissipating the electromagnetic waves energy. The low loss criteria is defined from propagation constant expression as (Annan, 2001):

$$\omega\mu\sigma \langle\langle \omega^2\varepsilon\mu \qquad (50)$$

or

$$\frac{\sigma}{\omega\varepsilon} \langle\langle 1 . \qquad (51)$$

**3.2.1.3 Concept of wavefronts and rays.** For more convenient characterization of electromagnetic waves propagation, the definition of wavefront and ray is introduced. In case of local source, the wavefront is defined as a surface of constant phase of transient signal at certain time after excitation (Figure 3.4). The ray is defined as a conceptual line along which the EM wave travels. The ray path is perpendicular to the wavefront and connects two points providing the shortest travel time.

In homogeneous media, wavefront surfaces are symmetrical relatively to the point source position. In inhomogeneous media, wavefront surfaces become asymmetric relatively to spatial changes in velocity of wave propagation.

In the mathematical calculations, the spherical wavefront, for convenience, is simulated as a superposition of planar wavefronts that are perpendicular to rays direction.

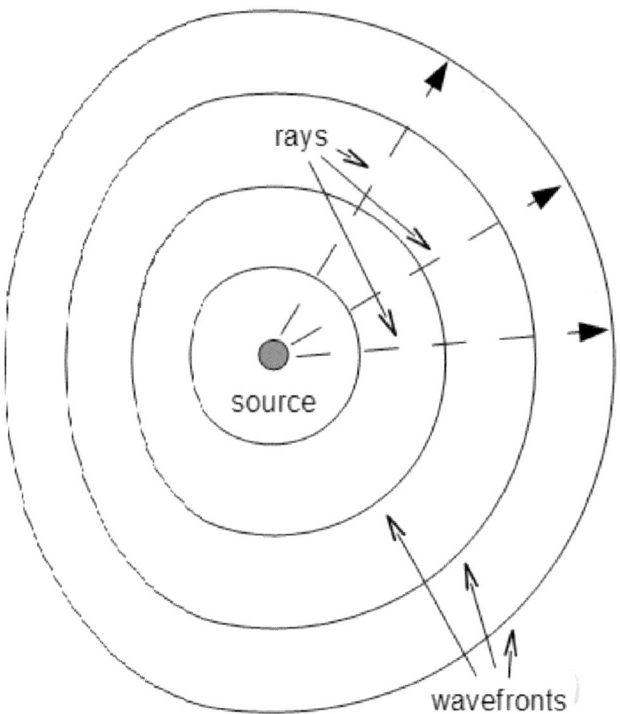

Figure 3.4. Wavefronts and rays for electromagnetic field of local source in

inhomogeneous media. Wavefronts are asymmetric because of difference in electric

properties (velocities of EM waves) of material.

**3.2.2. Plane Wave Reflection, Refraction and Transmission.** Predominantly, the GPR method maps objects using signals reflected from their surface. Reflections are created due to difference of electric properties of the object and surrounding materials.

**3.2.2.1 Reflection from the boundary.** At the boundaries, electromagnetic energy is partially reflected from and transmitted through the bounded media. For a plane interface between two materials of different electric properties (Figure 3.5), Snell's Law of reflection constitutes that

$$\sin \theta_i = \sin \theta_r, \qquad (52)$$

where

$\theta_i$ – angle of incidence

$\theta_r$ – angle of reflection.

It follows from (52) that $\theta_i = \theta_r$.

The Snell's Law of refraction relates the angle of incidence $\theta_i$ and refraction $\theta_t$ corresponding to transmitted signal:

$$k_1 \sin \theta_i = k_2 \sin \theta_t, \qquad (53)$$

where $k_1$ and $k_2$ are the wave-numbers of respective materials (see Figure 3.5).

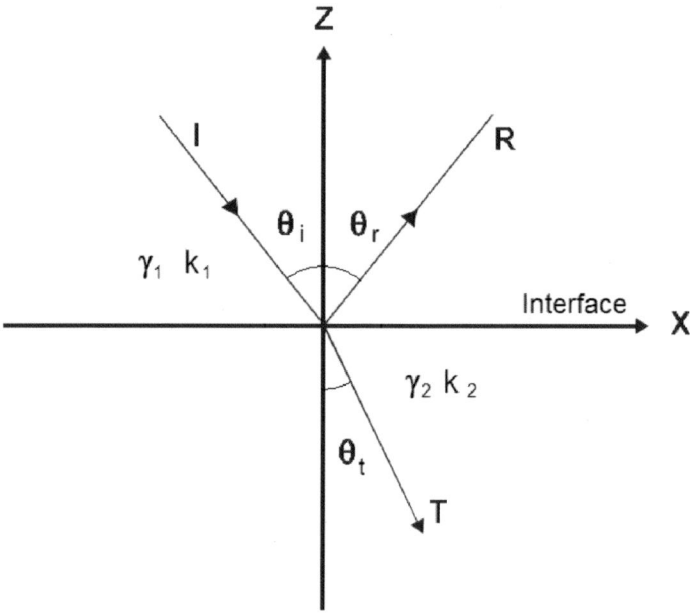

Figure 3.5. Diagram showing reflection and refraction of incident electromagnetic waves at planar interface of different materials. Incident signal (I) splits to reflected (R) and transmitted (T). Here k is a respective wave number.

For normal incidence ($\theta_i = \theta_r = \theta_t = 0$) of $y$-directed, $z$-polarized plane wave on the interface located on the $z$-$x$ plane, incident, reflected and transmitted phasor fields may be defined as

$$\mathbf{E}_S^I = E_{s0}e^{-\gamma_1 y}\mathbf{a}_z$$
$$\mathbf{H}_S^I = \frac{E_{s0}}{Z_1}e^{-\gamma_1 y}\mathbf{a}_x$$

for incident wave,

$$\mathbf{E}_S^R = RE_{S0}e^{\gamma_1 y}\mathbf{a}_z$$
$$\mathbf{H}_S^R = -R\frac{E_{S0}}{Z_1}e^{\gamma_1 y}\mathbf{a}_x \qquad \text{for reflected wave,}$$

$$\mathbf{E}_S^T = TE_{S0}e^{-\gamma_2 y}\mathbf{a}_z$$
$$\mathbf{H}_S^T = T\frac{E_{S0}}{Z_2}e^{-\gamma_2 y}\mathbf{a}_x \qquad \text{for transmitted wave.}$$

Here $R$ and $T$ are the reflection and transmission coefficients, respectively. Applying the boundary conditions that electric and magnetic fields must be continuous through the boundary (Hayt, 1989) one can write:

$$\begin{aligned} E_y^I + E_y^R &= E_y^T \\ H_x^I + H_x^R &= H_x^T \end{aligned} \qquad (54)$$

Solution of equations (54) gives the Fresnel reflection and transmission coefficients as

$$R = \frac{Z_2 - Z_1}{Z_2 + Z_1} \qquad (55)$$

$$T = \frac{2Z_2}{Z_2 + Z_1}. \qquad (56)$$

Using equation (49), one can find that at boundaries of high conductive media, e.g. metals, $R \approx -1$ and signal is reflected almost completely and the phase is changed by $\pi$.

In case of parallel polarization and arbitrary angle of incidence, the reflection and transmission coefficients can be found using Snell's Law for refraction (53) in the follow form:

$$R = \frac{Z_2 \cos\theta_t - Z_1 \cos\theta_i}{Z_2 \cos\theta_t + Z_1 \cos\theta_i} \qquad (57)$$

$$T = \frac{2Z_2 \cos\theta_i}{Z_2 \cos\theta_t + Z_1 \cos\theta_i}. \qquad (58)$$

Analysis of equation (57) shows that it is possible to get total transmission of energy at angle $\theta_B$ called *Brewster angle*, which can be defined for lossless media by

$$\theta_B = \sin^{-1}\sqrt{\frac{\varepsilon_2}{\varepsilon_1 + \varepsilon_2}}. \qquad (59)$$

Using Snell's Law of refraction (53), one can find that the electromagnetic energy is entirely reflected from boundary of material with higher velocity at the *critical angle* $\theta_c$ defined by

$$\sin\theta_c = \frac{k_1}{k_2} = \frac{v_1}{v_2}. \qquad (60)$$

**3.2.2.2 Reflectivity of thin layers.** Thin layers (thickness is less than wavelength) are the common targets of GPR investigation. The total reflected signal from a thin layer is formed by the superposition of the primary reflections from the boundaries and the multiples bouncing between interfaces with a phase shift depending on the thickness and velocity in the layer material (Figure 3.6).

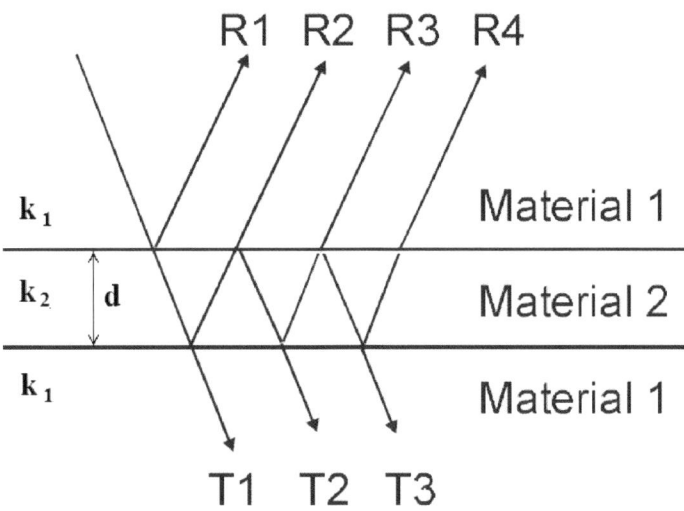

Figure 3.6. Reflection from the thin layer is formed by the primary reflections and the multiples bouncing between the interfaces (modified from Annan, 2001).

For normal incident sinusoidal wave the time delay $\tau$ between the superimposed signals can be defined as

$$\tau = e^{j2k_2 d} ,$$

where $k_2$ is a propagation vector in material of layer and $d$ is a layer thickness. Then the coefficient of reflection for a thin layer of material 1 surrounded by material 2 may be found as (Gottsche, 1997):

$$R = \frac{R_{12}\left(1 - e^{j\tau}\right)}{1 - R_{12}^2 e^{j\tau}} , \tag{61}$$

where

$\tau = 4\pi d/\lambda_2$ time delay

$\lambda_2$ - wavelength in material 2

$R_{12}$ – coefficient of reflection between materials 1 and 2.

For layer of thickness $d << \lambda_2$ the time delay $\tau << 1$ and equation (59) reduce to (Annan et al., 1988):

$$R \approx \frac{-R_{12} j\tau}{1 - R_{12}^2} . \tag{62}$$

One can find that the temporal form of signal reflected from thin layer is close to the time derivative of the incident wave (Annan, 2001). This suggests increasing the frequency of reflected pulse (Figure 3.7).

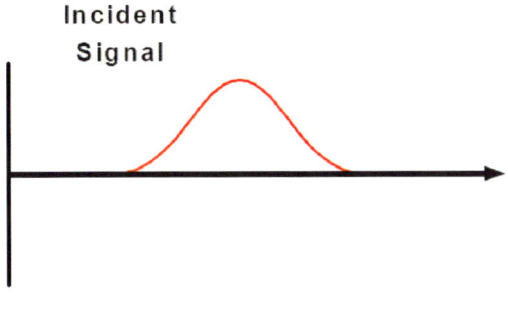

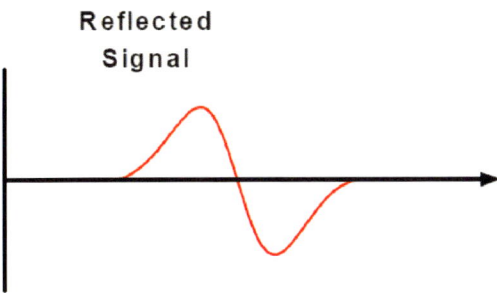

Figure 3.7. Simplified presentation of reflection from a thin layer. The resulting shape of

the signal becomes close to derivative of the incident pulse (from Annan, 2001)

Vertical resolution defines a minimum distance between two reflectors to be

distinguished on the GPR record. Theoretically, the minimum resolvable thickness of a

layer is one-quarter of wavelength (Sheriff and Geldart, 1995) or, more practically, a half

wavelength of a pulse, but thin layers are still detectable if the amplitude of reflected

signal exceeds the noise level (Annan, 1988).

**3.2.3. Electric Properties of Real Materials.** The GPR method is used to investigate subsurface employing electromagnetic waves propagating through an incredible variety of natural and manmade subsurface materials. The properties of these materials control the parameters of the electromagnetic waves spreading into the medium. It allows predicting the manner in which the radio waves penetrate the subsurface (if the properties of material are known) or defining the unknown material properties by inversion of GPR data.

In the most GPR applications, the electrical properties govern electromagnetic waves behavior in real media. The influence of magnetic properties is usually weak and is assumed negligible for most materials GPR deals with (Annan, 2001).

As described in above sections, the most important GPR parameters, such as velocity and attenuation, depend predominantly on the dielectric constant $\varepsilon_r$ and conductivity $\sigma$. These quantities determine the reflection coefficients because they form the complex electromagnetic impedance of the media. The electric properties of some typical geological materials are presented in Table 3.1.

In most cases, the GPR method deals with the media composed with a mixture of different materials. Therefore, understanding GPR data considerably depends on understanding the properties of mixtures of materials (Annan, 2001).

Electromagnetic response of the mixture of materials seldom corresponds to the summarized response of proportional volume fractions of the material components. For example, water content strongly affects the properties of propagating radar signal because the real part of water dielectric constant is 80 and that of the most of dry materials is 4-8 (Topp et al., 1980). The bulk conductivity of geological materials mainly depends on the pore water content (Figure 3.8). In case of mixture of materials, GPR deals with the effective bulk properties.

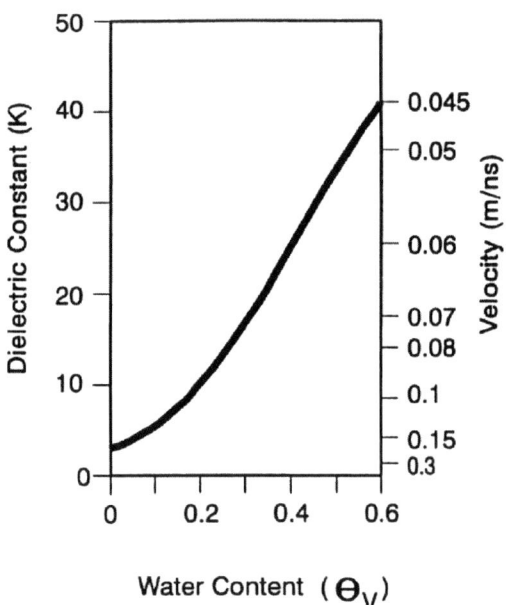

Figure 3.8. Variation of dielectric constant and velocity versus water content. As follows from the plot, dielectric constant of water saturated material is dominantly governed by water content (from Topp et al., 1980).

Another important factor influencing conductivity is surface conduction related to the charges trapped on the surface of mineral grains. Therefore, the fine-grained materials such as clay and silt having bigger surfaces per unit volume exhibit higher conductivity.

In Figure 3.9 a velocity versus frequency plot is shown. Velocities demonstrate a plateau at the 10-1000 MHz frequency range for the materials of conductivity of less than 100 *mS/m* and radio waves do not experience the dispersion caused by frequency-dependent velocity at these frequencies. Water molecule relaxation leads to an increase in velocity for frequencies above 1000 MHz.

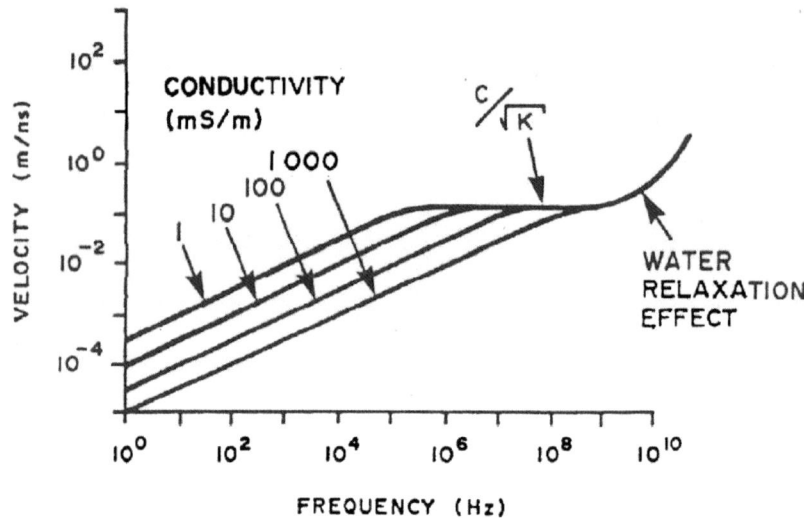

Figure 3.9. The relation between velocity and frequency at different conductivities (from Annan and Davis, 1989)

The plot of attenuation versus frequency at different conductivities for materials with a dielectric constant 4 (Figure 3.10) shows that attenuation is almost constant in the georadar frequency range (10-1000 MHz). A rapid increase of attenuation is observed at frequencies higher than 1000 MHz because of water relaxation. Because attenuation has no extended plateau at GPR frequency range it can be a main parameter contributing to dispersion of radio signal. One can find more detailed discussions about electrical properties of real materials in Olhoeft, 1987.

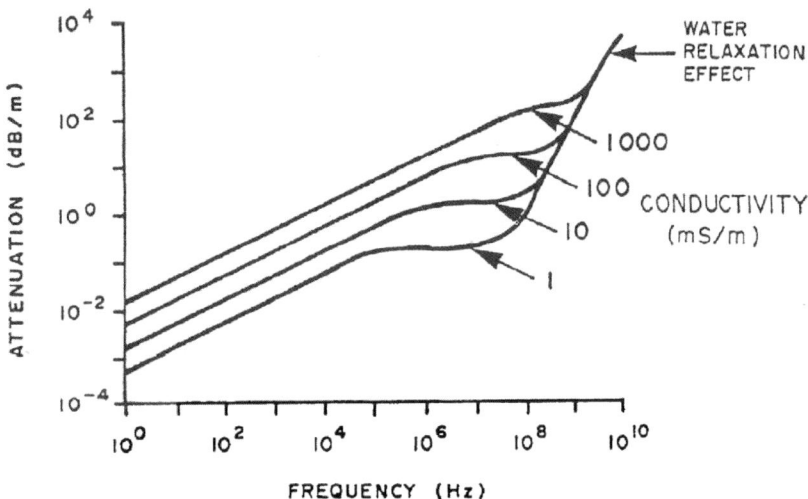

Figure 3.10. The relation between attenuation and frequency at different conductivities (from Annan and Davis, 1989)

There is limited information about electrical properties of salts and related geological rocks in publications. In Table 3.2 the dielectric constants and attenuation of salts and related materials obtained in the Saskatchewan potash mines are presented.

Theoretical reflection coefficients can be calculated using equation (55) for normal incidence of plane wave to the boundary between common materials (Table 3.3). The conductivity for these materials is assumed negligible.

Table 3.1. Typical electric properties of common geological materials (Davis and Annan, 1989).

| Material | Dielectric constant | Conductivity (mS/M) | Velocity (m/ns) | Attenuation (db/m) |
|---|---|---|---|---|
| Air | 1 | 0 | 0.30 | 0 |
| Distilled water | 80 | 0.01 | 0.033 | $2 \times 10^{-3}$ |
| Fresh water | 80 | 0.5 | 0.033 | 0.1 |
| Sea water | 80 | $3 \times 10^{4}$ | 0.01 | $10^{3}$ |
| Dry sand | 3-5 | 0.01 | 0.15 | 0.01 |
| Saturated sand | 20-30 | 0.1-1.0 | 0.06 | 0.03-0.3 |
| Limestone | 4-8 | 0.5-2 | 0.12 | 0.4-1 |
| Shales | 5-15 | 1-100 | 0.09 | 1-100 |
| Silts | 5-30 | 1-100 | 0.07 | 1-100 |
| Clays | 5-40 | 2-1000 | 0.06 | 1-300 |
| Granite | 4-6 | 0.01-1 | 0.13 | 0.01-1 |
| Dry salt | 5-6 | 0.01-1 | 0.13 | 0.01-1 |
| Ice | 3-4 | 0.01 | 0.16 | 0.01 |

Table 3.2. Electrical properties of salt and related materials that encounter in salt and

potash mines (Annan et al, 1988).

| Material | Dielectric constant | Attenuation db/m |
|---|---|---|
| Pure salt and salt domes | 6 to 7 | 0.004 to 0.02 |
| Saskatchewan halite | 5 | 1 to 2 |
| Saskatchewan sylvite | 5 to 5.5 | 0.4 to 1.5 |
| Saskatchewan clay seam | 5 to 6.5 | 2 to 3 |
| Air | 1 | 0 |
| Water | 80 | 0.01 to 100 |

Table 3.3. Examples of theoretical normal incidence reflection coefficients for plane interface between some common materials (Annan, 2001)

| Boundary | $\varepsilon_1$ | $\varepsilon_2$ | $Z_1$ | $Z_2$ | $R$ |
|---|---|---|---|---|---|
| Air – dry soil | 1 | 4 | 377 | 188 | -0.05 |
| Air – wet soil | 1 | 25 | 377 | 75 | -0.67 |
| Dry soil – wet soil | 4 | 25 | 188 | 75 | -0.43 |
| Dry soil - rock | 4 | 6 | 188 | 154 | -0.01 |
| Wet soil - rock | 25 | 6 | 75 | 154 | +0.34 |
| Ice - water | 3.2 | 81 | 210 | 42 | -0.67 |
| Moist soil - water | 9 | 81 | 126 | 42 | -0.5 |
| Moist soil - air | 9 | 1 | 126 | 377 | +0.5 |
| Soil - metal | 9 | $\infty$ | 126 | 0 | -1 |

## 3.3. PRACTICAL ASPECTS OF GPR APPLICATION

This section is dedicated to the description practical aspects of the application of GPR method. GPR (ground penetrating radar, ground probing radar) is used to investigate subsurface objects whose electric properties differ from those of surrounding medium. Parameters of either reflections from interfaces or transmitted signals are employed to study the subsurface objects. The reflection method is commonly used for probing geological structures and is of most concern in this work. Despite a large number of successful GPR applications, there are some principal assumptions such as dielectric (lossless, low-loss) nature of geological materials, non-magnetic properties of the medium ($\mu_r=1$), frequency range (10 – 2500 MHz) providing domination of the displacement currents and minimum dispersion, that must be taken into consideration for a better understanding GPR data.

**3.3.1. Georadar Instrumentation.** GPR instruments emit EM energy into the ground, efficiently receive arrivals, digitize the received signal, store the digital data, and display the output radargrams. Some of the systems, supplied with computers, allow for basic processing and the printing of hardcopies. The general scheme of GPR system is depicted in Figure 3.11.

Basically, GPR instrumentation includes: antennas providing for the emission and receiving of EM energy; a control unit governing all the parameters of radiated signal, timing, amplifier and filter settings, and digitization rate; a laptop computer for handling the parameters of control unit, data storage and visualization. Usually, the transmitter, amplifier and digitizer electronics are combined with the antennas in separate blocks to reduce the noise generated in connecting cables.

**3.3.2. Antennas Characteristics.** Antennas are the most important elements of GPR instrumentation. Antennas define the central operating frequency, bandwidth of the pulse, and efficiency of EM energy emitting and receiving. The majority of the commercially available GPR systems use half-wave dipole antennas. From the term "half-wave", one can infer that the central frequency of dipole antenna depends on its length. Two types of antenna design are commonly employed: *monostatic* antenna using a single dipole for emitting and receiving the signal; and *bistatic* antenna using separate dipoles for emitting and receiving the EM energy.

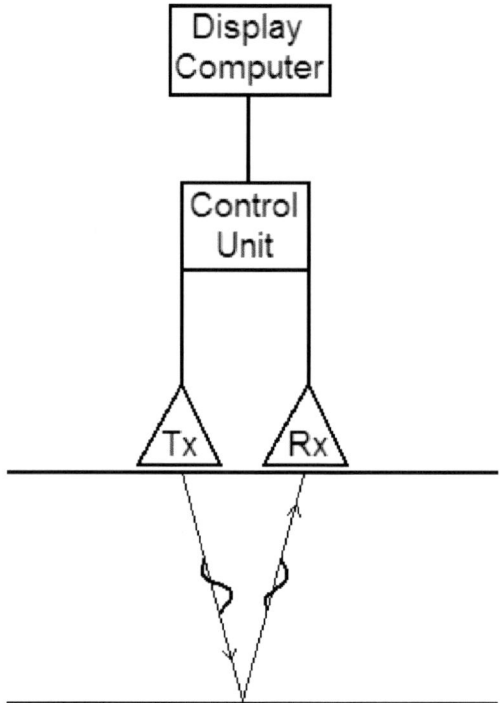

Figure 3.11. General scheme of GPR system. The EM signal is emitted by the transmitter antenna (Tx), captured by the receiver antenna (Rx), amplified, digitized, and stored.

**3.3.2.1 Antenna polarization.** The half-wave, center-fed, dipole antenna generates linear polarized electromagnetic waves with electric field component $E$ oriented parallel to the long axis of the dipole and magnetic field component $H$ oriented in the orthogonal plane. Based on the orientation of dipole relative to the vertical incidence plane (or survey line direction), three common types of the EM waves polarization are defined (Baker et al., 2007). Three types of dipole orientations relative to

the survey line for the case of bistatic antenna are shown in Figure 3.12.

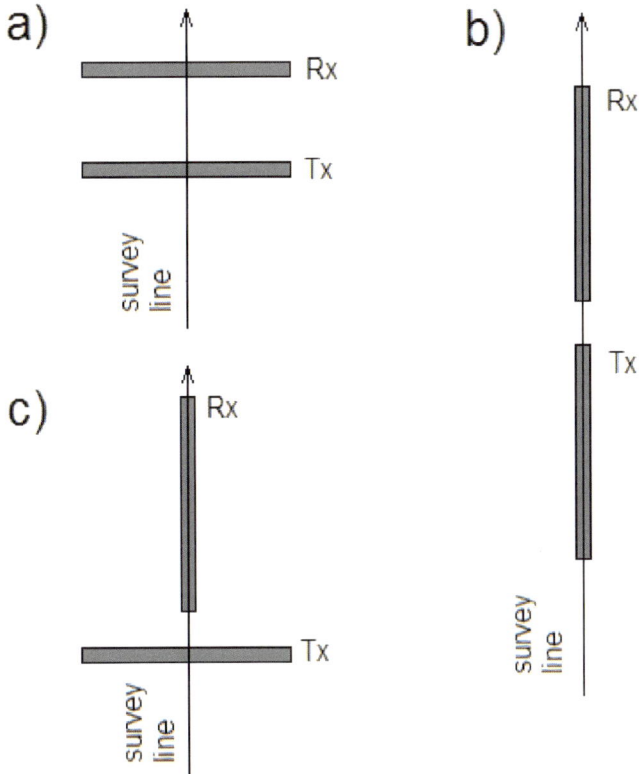

Figure 3.12. Three types of common antenna orientation: a) Perpendicular broadside

orientation (EH polarization, horizontal); b) Crossed dipole orientation (EH-EV

polarization, mixed); and c) Parallel end-fire orientation (EV polarization).

Perpendicular broadside orientation, when dipoles are oriented perpendicular to the survey line, is related to horizontal, perpendicular, E-horizontal (EH), or transverse electric (TE) polarization (Figure 3.12a). When the dipoles are oriented in the direction of survey line, the polarization is referred to as vertical, parallel, E-vertical (EV), or transverse magnetic (TH) (Figure 3.12c). The crossed dipole orientation provides with mixed EH-EV polarization (Figure 3.12b).

**3.3.2.2 Wave field of dipole on the surface of half-space.** Total-field solutions for a dipole in a lossless half-space provide insight into the different contributions depicted in Figure 3.13.

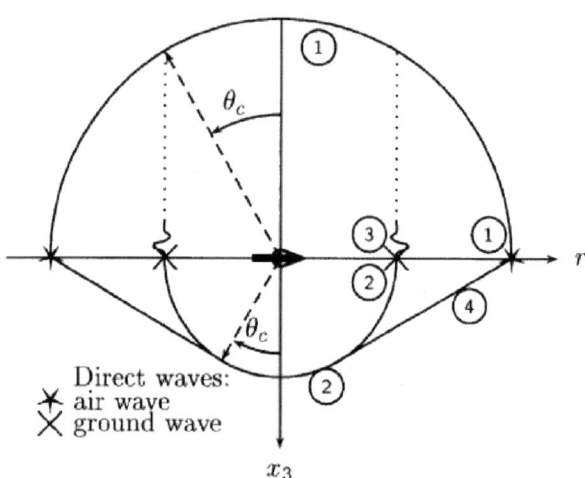

Figure 3.13. Wave field of the dipole on the surface of dielectric half-space. (1) Air body wave. (2) Ground body wave. (3) Inhomogeneous wave in air. (4) Ground head wave. Critical angle $\theta_c = sin^{-1}(1/\varepsilon_r^{1/2})$ (from Van der Kruk et al., 2003).

**3.3.2.3 Antenna directivity.** Another important characteristic of GPR is the directivity pattern. One of the GPR assumptions used for data processing and interpretation is the two-dimensional model of studied media. Actually, the dipole antennas generate three-dimensional electromagnetic fields that can lead to incorrect location of the target or a misunderstanding the reflections on the radargram. The simplified directivity pattern of the dipole is depicted in Figure 3.14.

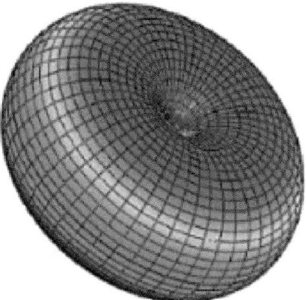

Figure 3.14. Directivity pattern of simple dipole in homogeneous medium (from Daniels, 2009).

The radiation of small electric dipole in spherical coordinate system (Figure 3.15) can be described by (Daniels, 2009)

$$E_r = \frac{2Z_0 Idl\pi \cos\theta_z}{\lambda^2} \left[ \left( \frac{\lambda}{2\pi r} \right)^3 \cos\psi + \left( \frac{\lambda}{2\pi r} \right)^2 \sin\psi \right] \tag{63}$$

$$E_\theta = \frac{Z_0 Idl\pi \sin\theta_z}{\lambda^2}\left[-\left(\frac{\lambda}{2\pi r}\right)^3\cos\psi - \left(\frac{\lambda}{2\pi r}\right)^2\sin\psi + \left(\frac{\lambda}{2\pi r}\right)\cos\psi\right] \tag{64}$$

$$H_\phi = \frac{Idl\pi \sin\theta_z}{\lambda^2}\left[\left(\frac{\lambda}{2\pi r}\right)^2\sin\psi + \left(\frac{\lambda}{2\pi r}\right)\cos\psi\right] \tag{65}$$

where

$dl$ - length of the current element

$\psi = (2\pi\phi/\lambda) - \omega t$

$\omega$ - circular frequency

$t$ - time ($=1/f$)

$c$ - velocity in vacuum

$Z_0$ - free space impedance (= 377 $\Omega$)

$I$ - current in the element

$\theta_z$ - zenith angle to the radial distance $r$

$\lambda$ - the wavelength

$r$ - distance from the element to the observation point.

Directivity strongly depends on the properties of the media surrounding the antenna. For a GPR survey it is very important to know the electromagnetic field behavior when the dipole is placed on the ground surface. The directivity patterns of dipole antenna in a half space placed at the boundary of air and materials of various permittivity are shown in Figure 3.16.

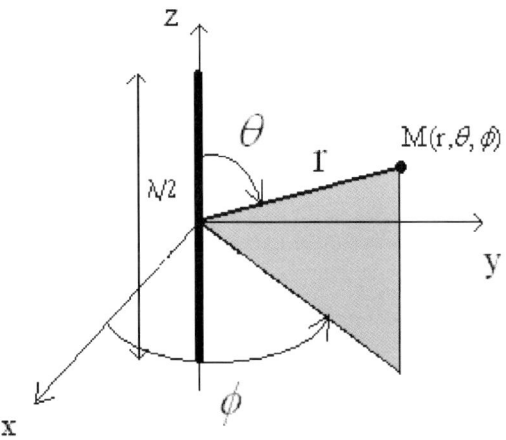

Figure 3.15. Spherical coordinate system used for definition of half-wave dipole

directivity pattern. Here M is an observation point. Dipole is oriented parallel to z-axis.

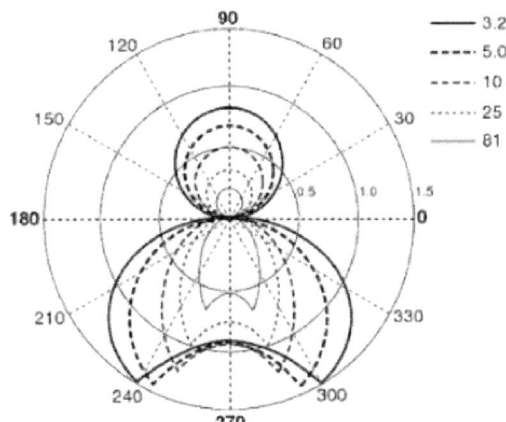

Figure 3.16. Transverse electric directivity pattern of dipole antenna on the ground

surface for various subsurface material permittivities. Material permittivity values are

indicated (from Annan, 2009).

**3.3.3. GPR Equation.** The GPR equation characterizes the radar system performance relative to technical parameters of transmitter, receiver, antenna characteristics, and properties of media and target object. The radar equation can be defined as (Davis and Annan, 1989)

$$Q = \frac{\xi_T \xi_R G_T G_R g S e^{-4\alpha L}}{64\pi^3 f^2 L^4} \tag{66}$$

where

$Q$ – system performance (ratio of the emitted signal amplitude to the minimum receiver sensitivity)

$\xi_T$ – transmitter antenna efficiency

$\xi_R$ – receiver antenna efficiency

$G_T$ – transmitter antenna gain

$G_R$ – receiver antenna gain

$L$ – distance to the target

$\alpha$ - attenuation coefficient

$f$ – frequency

$g$ – backscatter gain of target

$S$ – target area.

The technical parameters of instrumentation are the easy controlled factors that influence the system performance and can be measured enough precisely. Properties of the ground and target are most difficult to predict and getting a reliable estimate of radar signal range requires the trial testing at the site. Approximately, the signal range relatively to system performance and attenuation in material can be defined from the chart in Figure 3.17.

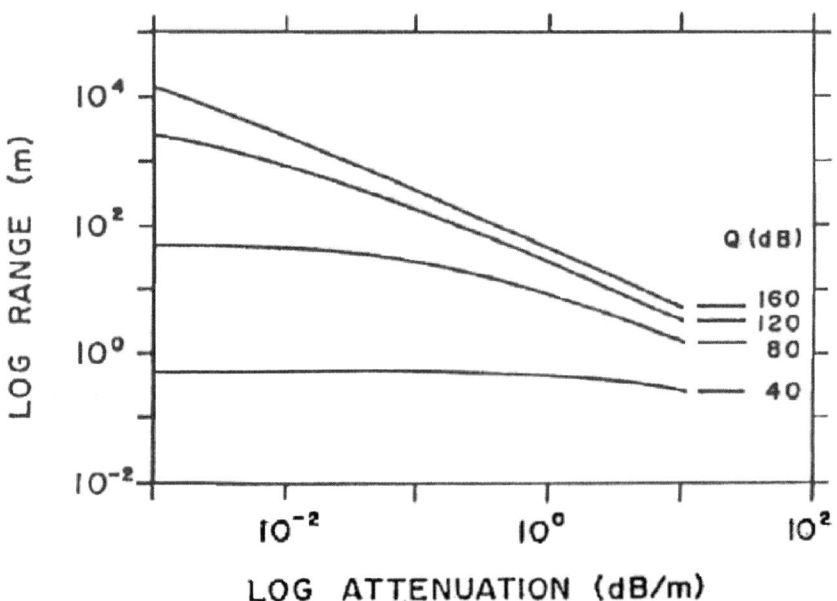

Figure 3.17. Relation between signal range and attenuation for various system performance (from Davis and Annan, 1989).

**3.3.4. Data Acquisition Modes.**  There are four commonly used GPR data acquisition modes that replicate the data acquisition modes routinely employed in the practice of seismic exploration (Sheriff and Geldart, 1995):

- continuous common offset profiling

- common midpoint sounding (CMP)

- common source sounding or wide-angle reflection and refraction (WARR), and

- radar tomography (transillumination, transmission) mode.

Each of these modes has specific instrumentation particularities and different

methodologies of data processing and interpretation, and is adapted to certain target

objects and environmental settings.

**3.3.4.1 Continuous common offset profiling mode.** The continuous common

offset profiling mode is the most often used mode in the practice of GPR surveying

because it provides for the acquisition of a large volume of data in relatively short period

of time and requires minimum personal effort.

The data are acquired effectively continuously by moving the monostatic or

bistatic (the transmitter-receiver offset is fixed) antennas along a traverse (Figure 3.18).

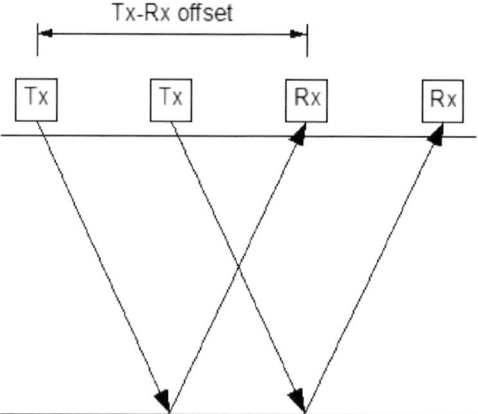

Figure 3.18. Simplified sketch of the continuous common offset profiling data acquisition

mode using bistatic antenna. Here Tx is a transmitter antenna position and Rx is a

position of receiver antenna.

The GPR section is a set of A-scans (point measurement records, traces) collected at uniform spacing (B-scan data). Interpretation of continuous common offset profiling data requires providing velocity information from other sources, e.g. CMP records, or it can be retrieved from the analyses of diffraction hyperbolas if available.

**3.3.4.2 Common midpoint sounding mode.** Also this acquisition mode is related to step mode when the traces are collected steadily by moving the transmitter and receiver antennas away from one another to keep the reflection point fixed at same position on the interface. The CMP data acquisition configuration is depicted in Figure 3.19.

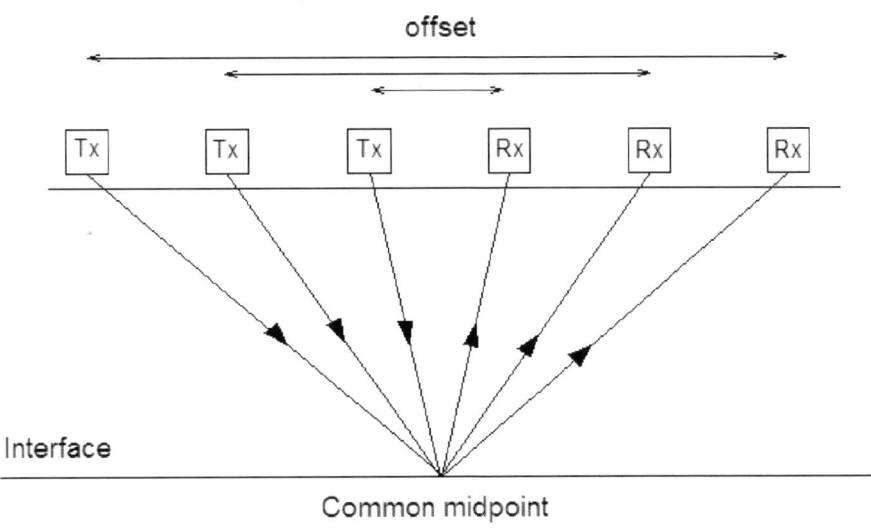

Figure 3.19. CMP sounding data acquisition configuration. The transmitter and receiver antennas are moved away to keep the same position of reflection point.

Common midpoint sounding mode configuration is mostly used for velocity evaluation purposes. The radargrams resulting from measurements can be processed using conventional methods of velocity analysis employed in seismic exploration, e.g. velocity semblance analysis (Sheriff and Geldart, 1995). Sometimes the CMP method is used for profiling to get high-resolution data or continuous velocity information about subsurface but data acquisition becomes time and labor consuming.

**3.3.4.3 Common source sounding mode.** Common source sounding mode data acquisition is used when the CMP sounding mode data are technically unavailable or the reflection boundary is flat and parallel to observation surface. In the common source sounding mode data acquisition configuration (also called wide-angle reflection and refraction sounding, WARR), the transmitter antenna is kept at fixed position and the receiver antenna is moved away at uniform station spacing. This mode is also related to step mode data collecting because each trace is recorded after stopping at each observation station. WARR mode is depicted in Figure 3.20. This configuration is generally used for velocity determination, but application of this system in practice is limited because of erroneous results in case of dipping or uneven interfaces.

**3.3.4.4 Transillumination mode.** In the transillumination mode, the transmitter and receiver antennas are placed on opposite sides of the studied media (space between the boreholes, mine pillars). Both transmitted and scattered signals are recorded and analyzed. The simplified chart of transmission mode configuration is shown in Figure 3.21.

The transillumination mode is commonly used when the access to the object is limited or signal penetration depth is not enough to image the target.

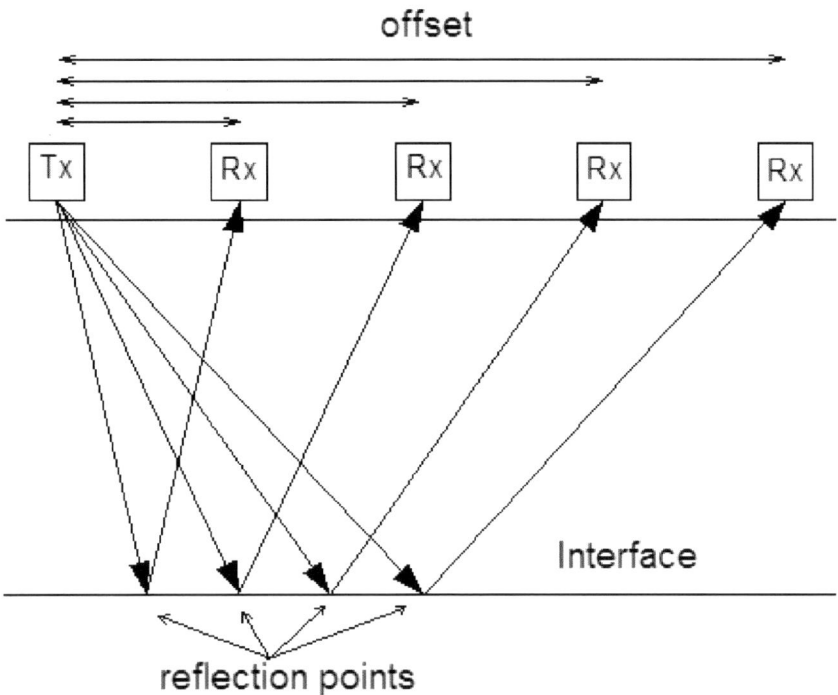

Figure 3.20. Common source data acquisition mode. The source is fixed and receiver is moved away.

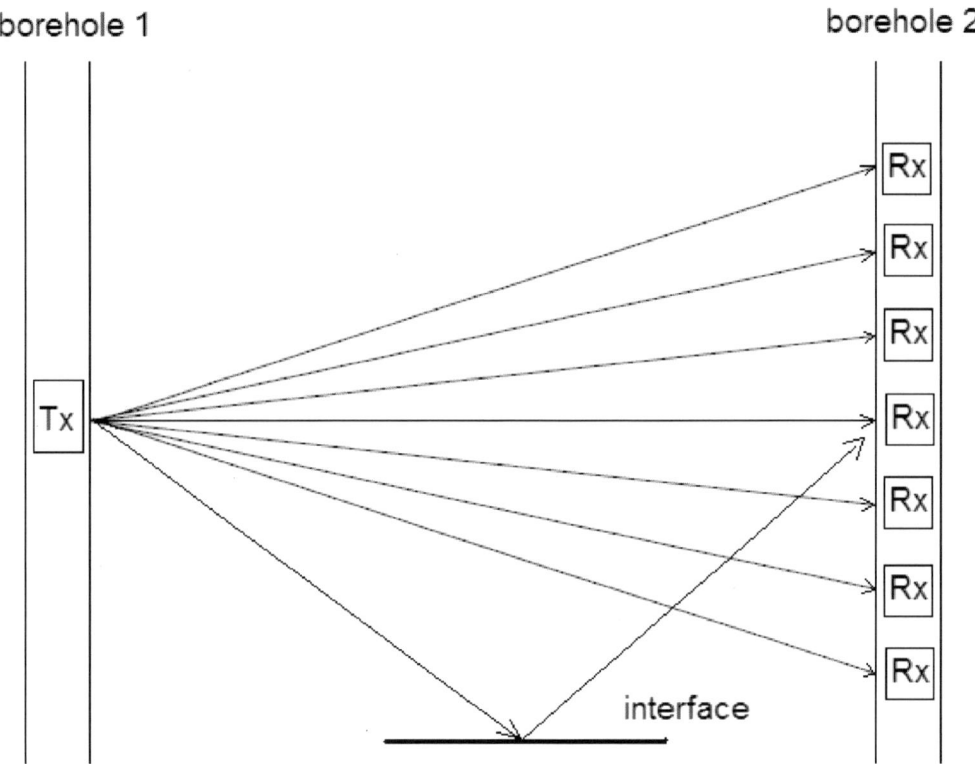

Figure 3.21. The cross-borehole transillumination configuration. The transmitter and receiver antennas move in adjacent boreholes to characterize the structure and properties of the media between them.

### 3.3.5. Resolution and Penetration Depth Problem.

The trade-off between resolution and penetration depth is one of the principal problems of the GPR method. Because attenuation in real geological materials is frequency dependent, the amplitude of higher frequency signal decays faster. On other hand, resolution is lower for lower frequency GPR signal.

Vertical resolution defines the minimum separation between two reflectors that still allows the reflectors to be distinguished on GPR records (Figure 3.22).

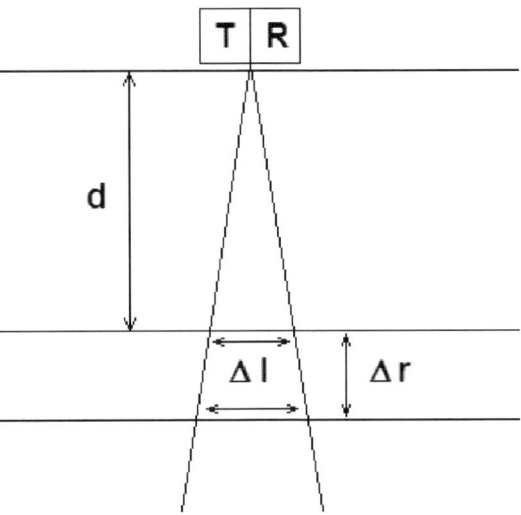

Figure 3.22. Vertical ($\Delta r$) and horizontal ($\Delta l$) resolution. Here $\Delta l$ – horizontal separation between two objects, $\Delta r$ – vertical separation between two interfaces, d – depth of interface.

Vertical resolution (longitudinal, range, depth resolution) defines a minimal separation between two interfaces oriented perpendicular to the direction of wave propagation. In publications, the theoretical minimum resolvable thickness of a layer is reported to vary from one-quarter (Sheriff and Geldart, 1995) to a half of wavelength (Annan, 2009). Note, that in Annan (2009), vertical resolution is indicated as the width of pulse equivalent to a half-period cycle. Vertical resolution values for a 0.12 m/ns velocity as a function of the central operating frequencies used for my experimental studies are shown in Table 3.4.

The lateral or horizontal resolution criterion is based on the premise that two targets are not distinguishable if the distance between them is less than size of first Fresnel zone. Consequently, horizontal resolution $\Delta l$ can be defined as (Annan, 2001)

$$\Delta l = \sqrt{\frac{d\lambda_c}{2}} \, , \tag{67}$$

where $\lambda_c$ is a wavelength of the central frequency of the GPR antenna.

Table 3.4. Vertical resolution parameters for electromagnetic waves in material with a velocity of 0.12 m/ns. The resolution is calculated using the one-quarter wavelength criterion.

| Frequency (MHz) | Wavelength (m) | Resolution (m) |
|:---:|:---:|:---:|
| 50 | 2.4 | 0.6 |
| 150 | 0.8 | 0.2 |
| 250 | 0.48 | 0.12 |
| 400 | 0.3 | 0.075 |
| 1200 | 0.1 | 0.025 |

**3.3.6. GPR Survey Sampling Fundamentals.** GPR tools digitally record electromagnetic field magnitudes in time and space. To ensure the digital signal correctly replicates the recorded field, the recorded data must satisfy Nyquist sampling criteria.

For a signal with maximum frequency $f$, the time $\Delta t$ and spatial $\Delta x$ intervals must be (Annan, 2009)

$$\Delta t \leq \frac{1}{2f} \tag{68}$$

$$\Delta r \leq \frac{v}{2f}. \tag{69}$$

For wideband transient GPR signals this criteria may be written as

$$\Delta t \leq \frac{1}{6f_c} \tag{70}$$

$$\Delta r \leq \frac{v}{6f_c}, \tag{71}$$

where $f_c$ is the central frequency.

# 4. DETERMINATION OF ELECTROMAGNETIC PROPERTIES OF SALT ROCKS

The effectiveness of GPR data interpretation depends on knowledge of the propagation velocity of electromagnetic waves in the medium. The propagation velocities of radio signals in salt rocks of the Upper Kama deposit were determined through laboratory samples testing, the analyses of common-midpoint (CMP) and wide-angle reflection and refraction (WARR) data acquired in the field, and the fitting diffraction hyperbolas into radargrams.

## 4.1. LABORATORY TESTING

Laboratory determination of the electromagnetic parameters of salt rocks was conducted in the Laboratory of Geomechanics in the Mining Institute of Ural Branch of the Russian Academy of Sciences. Measurements were conducted on salt rock samples extracted from a potash mine. The rock samples were small in size (~ 40-50 cm high and about 30 cm wide). Therefore, the most portable high-frequency 1200 MHz antenna was employed for measurement purposes. The simplest measurement setup available was the analysis of bottom reflection travel time because the transmitter and receiver elements of this antenna are housed in a single box. The velocity evaluation setup is depicted in Figure 4.1.

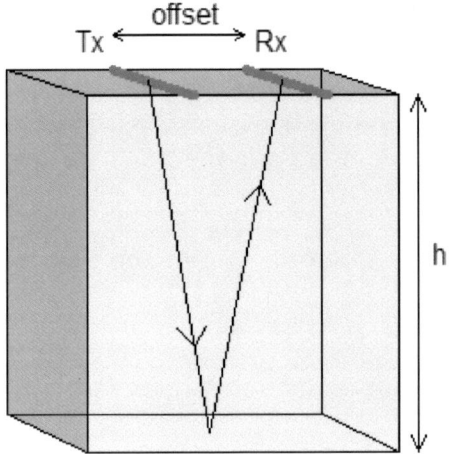

Figure 4.1. Measurement setup for velocity evaluation on salt rock samples. Offset

between transmitter and receiver elements is 6.1 cm. Here h is the height of the sample.

The velocity $v$ can be calculated using the travel time equation:

$$t(\textit{offset}) = 2\left(h^2 + \left(\frac{\textit{offset}}{2}\right)^2\right)^{\frac{1}{2}} \Big/ v, \qquad (72)$$

where *offset* is the distance between transmitter and receiver, *t(offset)* is the travel time of

reflected signal and *h* is the height of the sample. Using (72), the velocity may be defined

as

$$v = 2\left[ h^2 + \left( \frac{offset}{2} \right)^2 \right]^{1/2} \Big/ t(offset).$$
(73)

The offset of 1200 GHz antenna was 6.1 cm and the height of samples used for testing was 40-50 cm. Consequently, $(offset/2)^2 << h^2$ and equation (73) can be reduced to

$$v = \frac{2h}{t}.$$
(74)

Radargram probing the rock salt sample of 40 cm height is shown in Figure 4.2. The traces were collected at different locations on the surface of the sample and the resulting velocity was obtained by averaging the measured data. The signal reflected from the bottom of sample was well recognized and picked for velocity calculation. Simple processing including start time and gain correction was used to improve the interpretation of radargrams.

The velocities were found to be in range 0.12-0.13 m/ns. Examples of radargrams obtained on the sample of rock salt of 40 cm and 50.5 cm height are shown in Figure 4.2 and Figure 4.3 respectively.

Sylvinite samples of appropriate size were not available for testing but it is established by previous studies in potash mines that the velocity of sylvinite and rock salt are similar (see Table 3.2). Velocity evaluation in carnallite rock was conducted on the sample of 16.5 cm height. The measured velocity was 0.13 m/ns (Figure 4.4).

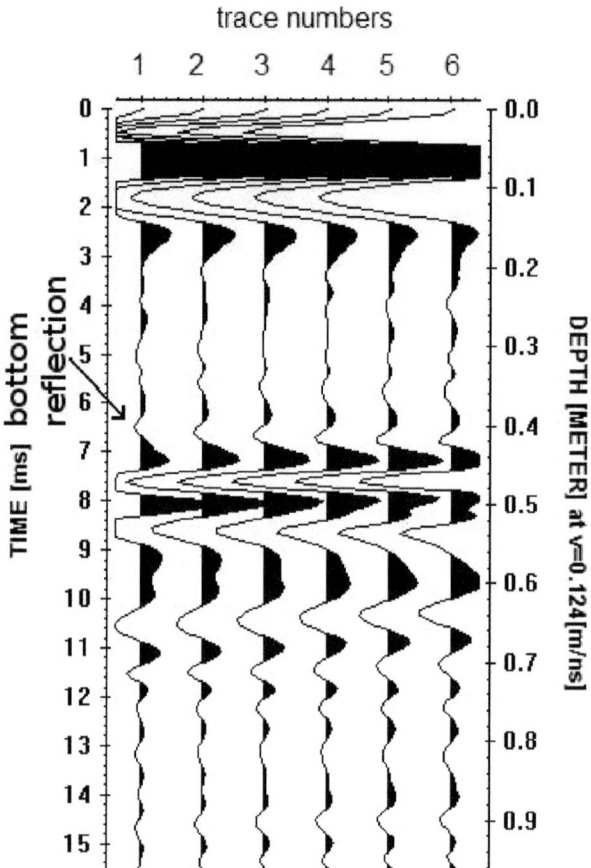

Figure 4.2. Radargram obtained on the sample of rock salt of height 40 cm using 1.2 GHz antenna. The depth scale calculated using velocity 0.124 m/ns shows that bottom reflection depth position is consistent with the known height of the sample. The bottom reflection has complicated shape due to superposition of reflections from base plate beneath the sample.

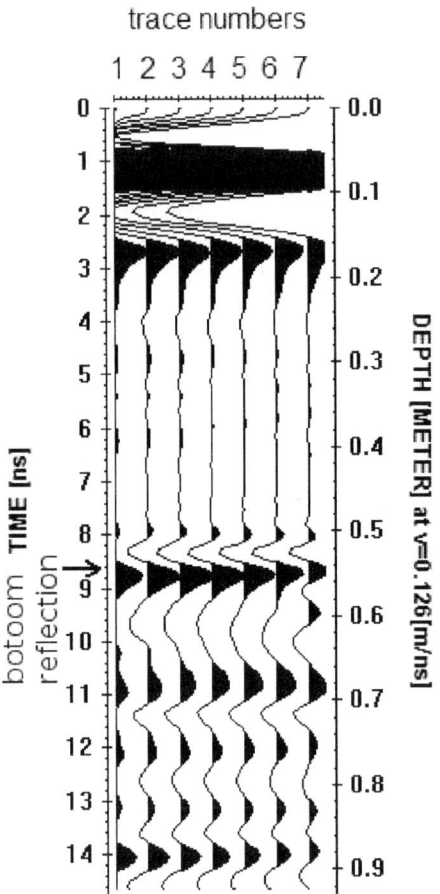

Figure 4.3. Radargram of testing the rock salt sample of 50.5 cm height. The measured average velocity is 0.126 m/ns.

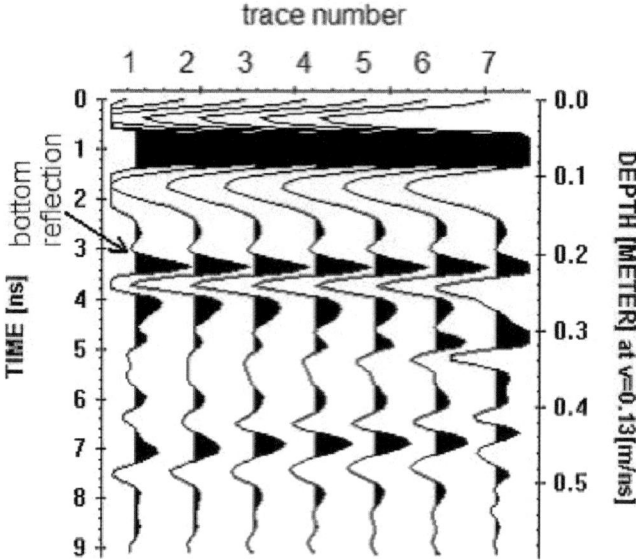

Figure 4.4. Radargram of probing the carnallite sample of 16.5 cm height. The estimated average velocity is 0.13 m/ns.

The higher velocity could be caused by less clay material content or lower accuracy of measurements because of small size of tested sample. The changes in velocities could be caused also by difference in stress state of the rock mass investigated (Orlando et al., 2010).

## 4.2. FIELD TESTING

CMP and WARR measurements were conducted in the sylvinite zone and in the underlying rock salt to estimate the *in situ* properties of the evaporite strata. Common mid-point testing was conducted in chamber 166 of the AB mining level. Walk-away sounding of the mine pillar of about 6 m thick was carried out to obtain reliable reflections from opposite wall. Separate transmitter and receiver antennas of central frequency of 250 MHz (OKO GPR system) were used to record the data at spacing interval of 0.5 m, with minimum offset of 0.5 m and maximum offset of 5 m. The semblance velocity analysis of CMP radargram obtained at the mine pillar consisted mainly of sylvinite rock demonstrate the approximately uniform velocity 0.12 m/ns (Figure 4.5).

The WARR (Wide Angle Reflection and Refraction) test was conducted in the conveyor drift driven in the underlying rock salt 20 m below the level AB. Data were acquired using RAMAC/GPR system with 50 MHz antennas. The receiver antenna was moved away from fixed transmitter at the interval of stations of 1 m with minimum offset of 1 m and maximum offset 46 m. The semblance analysis showed the very uniform velocities of electromagnetic waves propagation in the rock salt (Figure 4.6).

The reflection from the Marker Clay layer yielded the velocity of 0.118 m/ns. The relatively low velocity value may be explained by more significant contribution of bulk clay content depended on the propagation distance and, probably, by folding of clay layer.

Analysis of CMP radargrams and diffraction hyperbolas on radar sections showed that, in most instances, an average velocity of 0.12 m/ns could be used for interpretation and time-to-depth transformation.

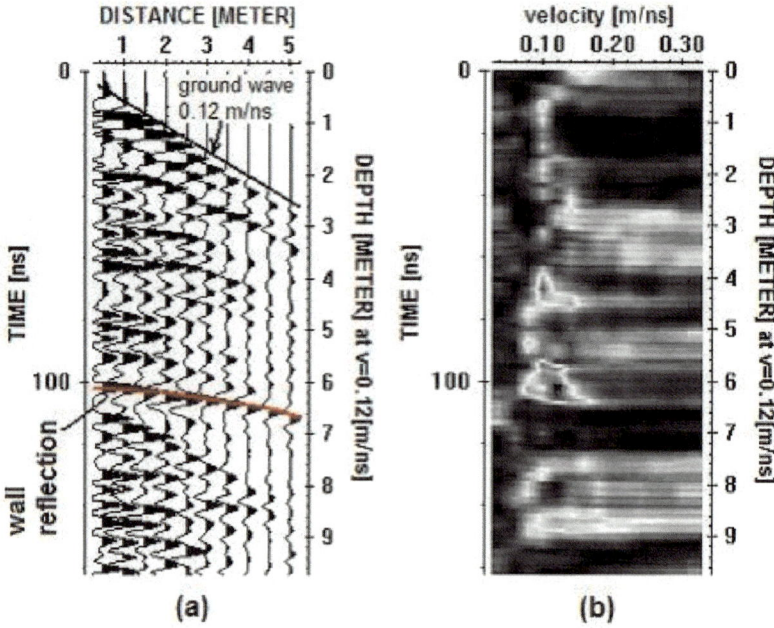

Figure 4.5. Semblance (b) panel of CMP data (a) obtained at the pillar of about 6 m thick. Start-time and Automatic Gain (AGC) corrections were applied to the data. The reflection from opposite wall is well recognized. The determined velocity (0.12 m/ns) is approximately uniform across the pillar. The first arrivals are presented by direct ground wave of velocity of 0.12 m/ns. Direct air wave is considerably reduced by antennas shielding.

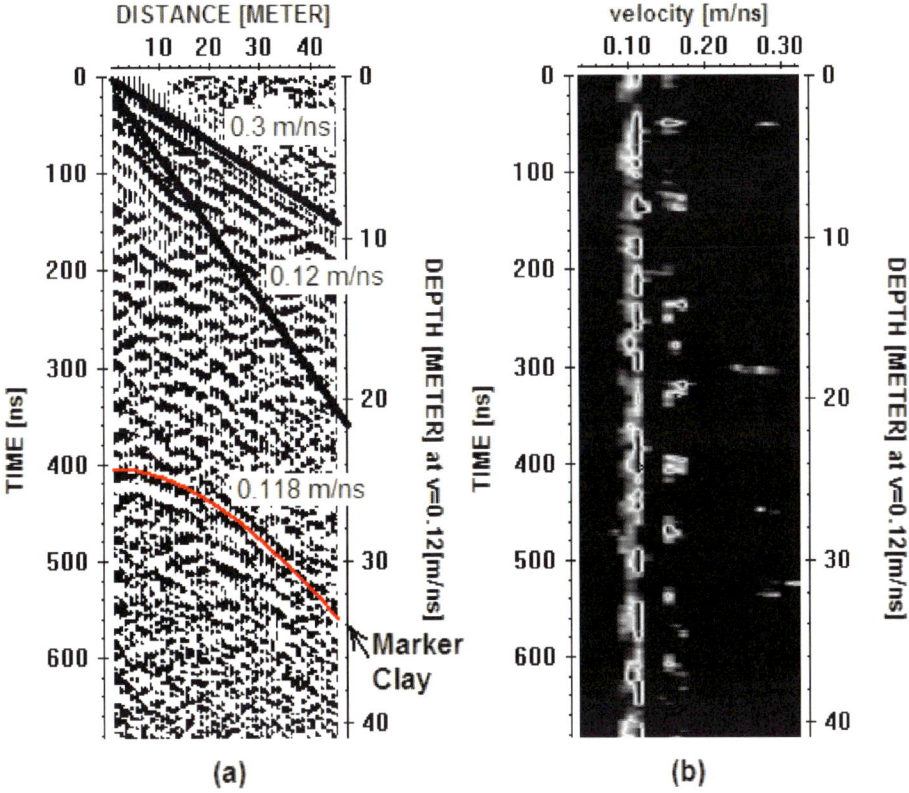

Figure 4.6. Semblance (b) panel of WARR data (a) obtained in the rock salt beds. The data were corrected for start-time delay and gain (AGC). Very uniform velocity of underlying rock salt (0.118 m/ns) was observed. The used 50 MHz antennas are not shielded and the first arrivals are presented by air wave of velocity of 0.3 m/ns. The ground wave of velocity of 0.12 m/ns also is clearly observed.

## 4.3. VELOCITY ANALYSIS USING DIFFRACTION PATTERNS

The average propagation velocity can be determined using the diffraction patterns commonly presented in the georadar sections. The GPR data processing software usually allows for the interactive fitting the synthetic hyperbolas onto the real data. The velocity

corresponding to the hyperbola best matching the diffraction pattern on the data image is

assigned to the material. A two-dimensional velocity model can be constructed if a

number of diffraction patterns are available throughout the radargram. An example of

hyperbola fitting onto the GPR data acquired along the mine pillar is presented in Figure

4.7.

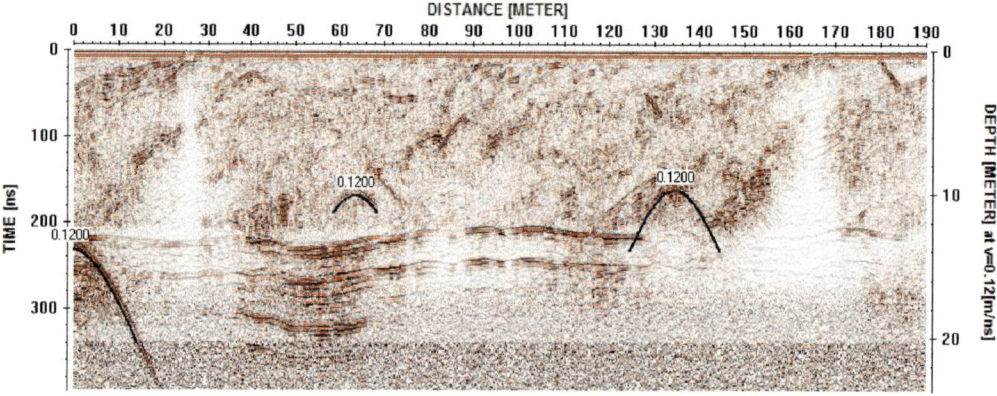

Figure 4.7. Velocity analysis of the radargram obtained along a mine pillar with a

thickness of about 13 m. Data were acquired using 250 MHz antennas. Synthetic

hyperbolae with a velocity of 0.12 m/ns best match the diffraction patterns throughout the

profile. The excellent diffraction hyperbola at the lower left part of the image is generated

by the corner of the pillar.

The bulk clay content mainly controls velocity variations in the salt rocks of the

Upper Kama potash deposit. Locally, an increase of velocity caused by fracturing (because of higher velocity in air) can be observed. A decrease in the velocity in the brine-saturated rock is expected. No special studies of velocity variation in the Upper Kama potash mines were conducted. Recommendations for further work on the determination of variation of the electromagnetic waves propagation parameters into salt rocks of the Upper Kama potash deposit are presented in the concluding Section.

**PAPER**

# 1. MAPPING OF EVAPORITE DEFORMATION IN A POTASH MINE USING GROUND PENETRATING RADAR: UPPER KAMA DEPOSIT, RUSSIA

## 1.1. ABSTRACT

Understanding the deformation processes in potash mines is very important for safe mining, planning the methods of extracting the orebody, and the prevention of catastrophic water inflow. A variety of deformational structures are present in the Upper Kama potash deposit. Folding is a dominant and most common form of deformation of ductile evaporites (herein also referred to as salts). Brittle deformation occur rarely but is of more significance from the perspective of mine safety because fractures affect the supporting capability of mine pillars and can provide passageway for water inflow. Continuous common offset ground penetrating radar (GPR) data were acquired in the potash mine operated by the Joint Stock Company (JSC) "Silvinit" to investigate a set of pre-existing fractures and related deformation structures. Open fractures, faults and folds were mapped using 2-D and 3-D GPR techniques. Detection of millimetric scale fractures was available due to high salt/air dielectric properties contrast and low background noise within uniform salt rock. FK filtering significantly improved fracture detection and imaging. A spatial model of one set of fractures was created using 3-D GPR imaging. Migration was applied to GPR data to obtain the true geometry of the strongly folded salt beds. The results of studies showed that GPR is capable of providing valuable information about deformation within the evaporite formations of the Upper Kama potash deposit.

*Keywords:* Ground penetrating radar, Evaporites, Fracture, Fold, Fault, GPR data processing

## 1.2. INTRODUCTION

A variety of natural and mine induced ductile and brittle deformation structures occur in the salt rocks of the Upper Kama potash deposit (Figure 1.1). Folding is the most common form of deformation encountered in the ductile evaporites (carnallite, sylvinite and halite herein referred to as salt). Natural brittle deformation is not commonly observed within the evaporitic strata of the Upper Kama deposit, but is of most importance for safe mining (Kudryashov et al., 2004; Jeremic, 1995). Studying the deformation in salt mines is required for safe operation, planning of extraction method, and preventing the catastrophic groundwater influx into mine openings.

Drilling is restricted in the salt mines because of economic reasons and fear of damaging the overlying water protective strata. Therefore, non-invasive geophysical methods are widely used to identify the potentially hazardous areas (Chouteau et al., 1997; Gendzwill and Stead, 1992; Neal et al., 1995; Thoma et al., 2003; Yaramanci, 2000).

The ground penetrating radar method (GPR) has proved to be very effective high-resolution tool, with a wide set of applications to geology, engineering, archeology, construction industry, and mining (Annan, 2002). The successful application GPR to fracture detection and study of deformation in crystalline and sedimentary rocks is reported in numerous publications (Grandjean, 1996; Busby, 1999; Orlando 2003; Apel and Dezelic, 2005a; Porsani et al., 2006). Toshioka et al. (1995), Grasmueck (1996) and

Lane (2000) described the usage of GPR for detecting and mapping millimetric cracks. A number of studies have demonstrated the enhanced capability of 3-D GPR imaging to delineate the subsurface structures (Young et al., 1997; Grasmueck et al., 2004; Gross et al., 2003; Christie et al., 2009).

The early GPR experiments in the salt and coal mines are described by Cook (1969) and Coon et al. (1981). The first experiments conducted in salt mines demonstrated that high-frequency electromagnetic waves are able to penetrate substantial distance in salt (Cook, 1969; Holzer at al., 1972; Stewart and Unterberger, 1976). Commonly, the GPR method has been used in salt mines to map stratigraphy, estimate the thickness of water protective beds, characterize fractures, detect unstable roof rocks, and evaluate the integrity of supporting pillars (Kovin, 2002; Thoma et al., 2003; Kelly et al., 2005). Annan et al. (1988), Gregoire and Halleux (2002), and Kovin (2010) describe the successful detection of fractures in the potash mines.

Preliminary experiments in the Upper Kama potash mines showed that the GPR method is capable of providing the detail and continuous information about rock mass structures at the distance of up to 40 m from mine openings (Kovin et al., 2002). As a non-destructive, portable and cost-effective method, the GPR method appears well suited for usage in the salt mine environment. Although the GPR method has been extensively used in German and Canadian salt mines for tens of years (Band et al., 1988; Annan et al., 1988; Chouteau et al., 1997), studies have shown that the GPR acquisition, data processing, and interpretation methodologies require an adaptation to local geological and mining environment.

This paper describes the results of the GPR studies of natural deformation structures observed within evaporitic formations of the Upper Kama deposit. The experimental field work was conducted in 2005 in order to evaluate the feasibility of using GPR data to detect and delineate fractures, and to map other deformation structures in salt rocks. Discussions of mine-induced deformation processes per se are beyond scope of this paper. The methodological aspects of GPR application in potash mines of the Upper Kama deposit are also presented.

## 1.3. GEOLOGICAL AND MINING SETTING

**1.3.1. General Geologic Setting of the Area.** The Upper Kama (Verkhnekamskoye) potash deposit is located in the northern part of the Perm kray (Russia), about 250 km north of the city of Perm (Figure 1.1). The Joint Stock Company (JSC) "Uralkali", headquartered in city of Berezniki, operates the southern part of the potash deposit, and the Joint Stock Company "Silvinit", headquartered in Solikamsk, operates the northern area.

The Upper Kama is second in size to Saskatchewan (Canada) Prairie Evaporite potash deposits among the world's currently mined potash occurrences (Garrett, 1995). The Upper Kama potash deposit covers an area of 3 700 km$^2$ (Kudryashov, 2001). Potash deposit consists of the potash-bearing beds located at the depths between 75-450 m (Garrett, 1995).

The term potash denotes a variety of mined and manufactured salts, all containing the element potassium in water-soluble form. Three minerals: sylvite (potassium chloride - KCl), carnallite (hydrated potassium magnesium chloride – KMgCl$_3$ 6H$_2$O) and halite (sodium chloride – NaCl) with subordinated amounts of anhydrite and clay typically form a potash ore. Sylvinite, as the most sylvite-rich crude ore, is mined in the Upper Kama deposit. There are three distinguished types of sylvinite ore (referred to as *red*, *banded* and *multi-colored* sylvinite), which differ in texture and sylvite content. Also carnallite is mined for the extraction of magnesium, as well as some amount of halite is mined for table and technical salt.

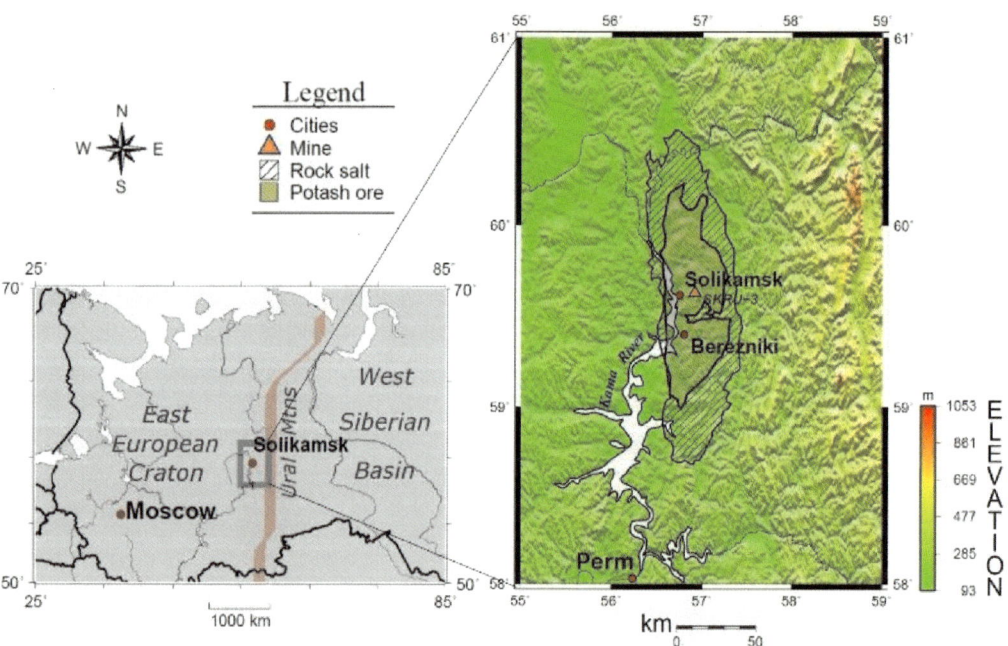

Figure 1.1. Location map of the Upper Kama potash deposit.

The Upper Kama potash deposit lies in central part of the Solikamsk (Solikamskaya) depression of the Pre-Ural foreland basin. This Permian (Kungurian stage) horizontally bedded evaporitic formation, up to 500 m thick, is composed predominantly of halite with substantial interbedded sylvinite, carnallite, clay, and anhydrite layers. The evaporite salts were deposited after the rise of the Ural Mountains in the late Carboniferous as a result of the continental collision between the East-European platform and West-Siberian plate (Rodgers, 1990; Zonenshain et al., 1990; Friberg et al., 2002).

The horizontally bedded evaporite formation is underlain by 2-3 km of clastic Proterozoic rocks, a Devonian reef sequence, a dominantly carbonate and carbonate-anhydrite Carboniferous sequence, and Lower Permian sediments up to 1 km in thickness. The Lower Permian molasse (flysch) wedge, consisting of conglomerate, sandstone and argillite beds, underlies the eastern part of the potash deposit (Figure 1.2).

The uppermost 80-120 m of evaporite formation contains economic deposits of potash salt, which include 13 sylvite and carnallite mineable beds and interbed halite layers (Garrett, 1995). Four sylvite-rich beds (referred to as A, Red I, Red II and Red III) in the lowest part of productive strata are the principal mineable beds (Figure 1.3).

The average thickness of a sylvinite zone is 17.4 m ranges from 3.3 to 30 m. Nine carnallite beds (namely B, V, G, D, E, Zh, Z, I, K) along with halite interlayers lie on top of the *banded* sylvinite layer A and form a carnallite zone that ranges in thickness from 38 to 80 m (Kudryashov, 2001). Locally, the sylvinite beds are replaced by barren halite and a host rock of carnallite beds is replaced by the *multi-colored* sylvinite or halite. The

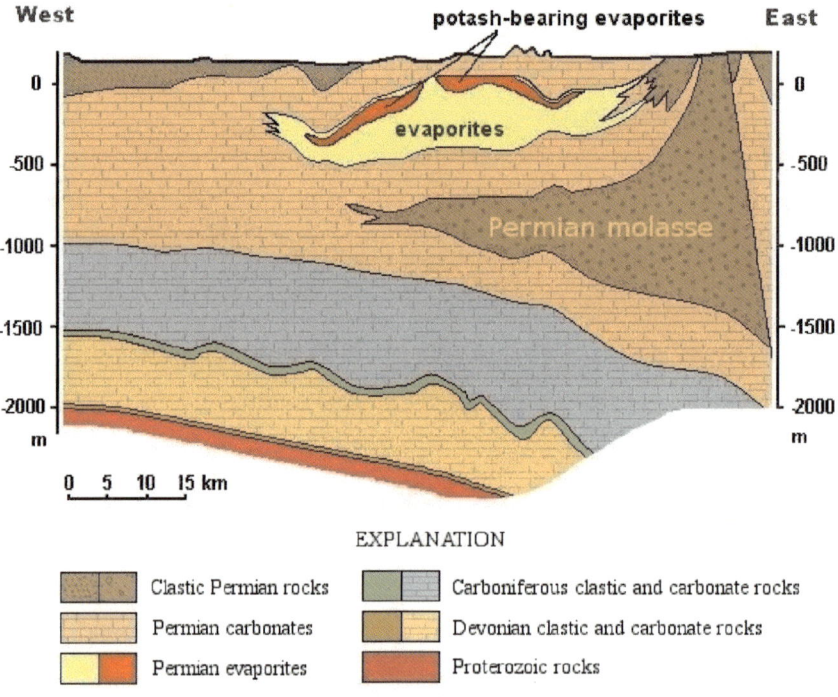

Figure 1.2. Generalized geological latitudinal cross-section of the Pre-Ural foreland basin and the Upper Kama potash deposit.

bonded A and B layers are usually mined together and are referred to as AB bed. Halite layer of 20 m thick, termed *overlying rock salt*, lies on a top of the potash beds.

Interbedded halite, clayey marl and clay layers of the *transition sequence* of approximately 20 m thick follow it. Clay, marl, limestone, sandstone, and siltstone (aleurolite) beds with multiple aquifer layers of above 200 m thick form the Middle Permian *terrigeneous-carbonate complex* overlying evaporites.

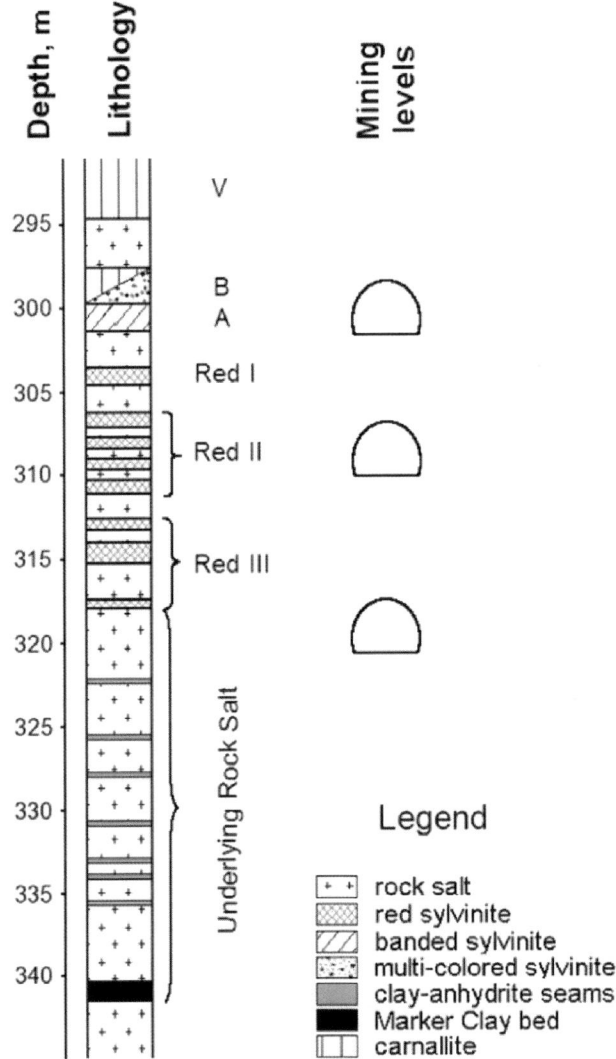

Figure 1.3. Lithological column of economic potash members in the study area with location of main mining levels. The mining levels AB and Red II are used for extraction the potash ore and the drifts at underlying rock salt are used for ore transportation and ventilation purposes.

Insolubles are presented by thin centimeter scale parting seams composed of clay and anhydrite. The 1 - 2 m thick carbonate clay layer (referred to as "Marker Clay") located in the underlying rock salt 20 m below sylvinite zone is the most stable stratigraphic marker at the Upper Kama deposit (Kudryashov et al., 2004).

The upper part of the potash deposit along with the overlaying marl, halite, clay, anhydrite, gypsum and carbonate beds of the transitional series is left un-mined to form a water protective barrier above the mine. The average thickness of impermeable strata is 70–90 m (Kovin, 2000).

**1.3.2. Forms of Deformation Structures in the Evaporitic Rocks.** The Upper Kama deposit has been subjected to major compressive regional tectonic, gravitational and mining induced forces. It was established that ductile evaporites react like a pseudo-fluid to the stress at geological time scale that results in ductile deformation of the evaporites rather than in brittle deformation (Brady and Brown, 2007; Warren, 1989).

Predominantly, tectonic movements in the adjacent orogens govern the folding deformation of the evaporite formations (Jeremic, 1994). The main folding structures of the Upper Kama potash deposit have submeridianal orientation, perpendicular to principal lateral E-W tectonic stress caused by the Uralian Orogen (Kudryashov, 2001). Locally, a superposition of a major tectonic stress and stress from the faults and numerous uplift structures leads to deformation reorientation or overlapping the deformational structures of different trend (Konstantinova et al., 2001).

Extensive folding of variable amplitudes and wavelengths is characteristic of evaporite formation of the Upper Kama deposit (Kudryashov, 2001). Westward inclined folds confirm the dominant East-to-West regional stress direction. The bending deformations with movement of beds along planes of stratification and flowage structures are commonly observed. The flow folding structures, often overturned, are met mostly in the sylvinite and carnallite layers.

Because of the high plasticity of the evaporitic salts, brittle deformation is not common in the Upper Kama potash deposit, but occurs locally when tectonic, gravitational or mine induced stress exceeds the strength of the rock mass (Lajtai et al., 1994; Kudryashov et al., 2004; Brady and Brown, 2007).

Geological and geophysical studies revealed systems of regional faults of longitudinal, latitudinal, northwest and northeast orientation (Kudryashov et al., 2004). The large solution structures, such as Durinskaya and Borovitskaya depressions, and collapse zone at Berezniki potash mine 3, are related to the faults in underlying and overburden sediments (Andrejchuk, 2002; Kudryashov et al., 2004).

Three regional meridianal thrust faults were established within the evaporite formation. However, the existence of some thrust faults is still debated because of difficult identification complicated by the intensive healing processes in salt rock (Kudryashov et al., 2004; Jeremic, 1995).

A number of discrete natural fractures and fracture systems of various scales have been observed during the operation of the potash mines, especially, in a central part of Upper Kama deposit. Most fractures are effectively healed and can be visually observed only in the non-annealed clay-anhydrite layers. Open fractures are encountered mostly in the sylvinite-carnallite zone. They usually develop in meridianal and NW-SE direction. Dominantly, the fractures have S-shape form and do not extend beyond the individual layer (Kudryashov et al., 2004). The most intense deformations are noted in the carnallite zone, where very steep folding, tectonic breccia and blocks of interbed rock salt are commonly observed (Kudryashov et al., 2004).

The largest set of identified fractures in the Upper Kama potash deposit was exposed at the eastern edge of mining field of the Solikamsk mine 3. Numerous oblique subvertical tension fractures are observed in the excavations of mining level AB at the total area 400 to 100 m. Opening of the striking NW 335°- 355° fractures ranges from less than millimeter to several centimeters. A few fractures were observed at the mining level Red II 8-9 m below level AB. It suggests that the depth of the fracture zone does not exceed 10 m. The upward exploration boreholes showed that the fractures die out at overlying carnallite layer or interbed interfaces. The fractures intersect the NE-SW trending folds and, probably, are resulted of local stress redistribution caused by NW-SE fault structure, which is located eastward from the site (Kudryashov et al., 2004).

### 1.3.3. Mining Setting and Problems of the Upper Kama Potash Mines.

Currently five underground mines produce a potash ore at the Upper Kama deposit. Two of them (BKRU-3 (Berezniki potash mine 3) and BKRU-1 (Berezniki potash mine 1)) were flooded in 1988 and 2006 respectively and abandoned. All the mines use a multiple-level room-and-pillar method of mining, and the rooms are 200 m length, 3m to 15 m width and 3m to 10 m high. Pillars between the rooms are of 3-18 m width. Depending on the number of extracted sylvinite layers, the mining levels are separated by 5-15 m of interbed rock salt.

In the Solikamsk mine 3, the potash ore is extracted from the depth of about 300 m. Two most sylvite-rich beds, AB and Red II, were mined at the survey site. The ventilation and conveyor tunnels were excavated under the potash layers in the underlying rock salt.

The water protective beds of about 130 m thick isolate the mine openings from overlying aquifers. Salt is a highly soluble material that makes preventing of inflow of water into a salt mine one of the most concern of mining engineers. Sudden failure of supporting pillars or reopening of natural fractures during mining operation can produce the pathways for unsaturated groundwater inflow into the mine causing a catastrophic collapse. According to mining regulations, the open fractures and high-amplitude overturned folds are supposed to be a prerequisite of fault structure and should be thoroughly investigated to mitigate the risk of massive roof rock failure and catastrophic flooding.

## 1.4. METHODOLOGY OF GPR INVESTIGATION IN A POTASH MINE

The application of GPR technique is based on propagation and reflection of high-frequency (10–2500 MHz) electromagnetic (EM) signals in the subsurface. Comprehensive description of the ground penetrating radar method is presented in Daniels (1996) and Annan (2001).

**1.4.1. Determination of Parameters of GPR Waves Propagation.** The velocity of propagation and attenuation of the signal are the main parameters used for the characterization of a material. Velocity of EM signal is governed by dielectric constant of the material. Attenuation of radar signal depends on the electrical conductivity and considerably reduces the penetration depth in conductive materials such as clay or mineralized brine. In resistive salts, substantial propagation distance of up to hundreds of meters can be achieved (Unterberger, 1978).

Velocity of propagation of a GPR signal was determined by laboratory testing of salt rock samples, and analyses of common-midpoint (CMP) and wide-angle reflection and refraction (WARR) data acquired in the mine, and fitting diffraction hyperbolas into radargrams. Studying of the salt rock samples was done in the Geomechanics Laboratory of the Mining Institute of Ural Branch of the Russian Academy of Sciences. The samples were relatively small in size (40-50 cm high and about 30 cm wide). Therefore, the most portable 1200 MHz antenna was employed for measurements. The velocity V was defined using simple two-way travel time relation $V=2d/t$, where d is a known height of sample and t is a measured travel time of signal reflected from a bottom of sample (Figure 1.4).

The velocities were found to be in the range 0.12-0.13 m/ns. The velocity obtained in the mine appeared slightly lower due to the bulk clay content. The CMP and WARR measurements were conducted within the sylvinite zone and in the underlying rock salt to estimate *in situ* the properties of evaporite rocks. Semblance velocity analysis of CMP radargram obtained at the mine pillar of approximately 6 m thick is shown in Figure 1.5.

Analysis of sounding radargrams and fitting of the diffraction hyperbolas into radar sections showed that, in most instances, an average velocity of 0.12 m/ns could be used for interpretation and time-to-depth transformation. This velocity is consistent with the results of GPR studies curried out at other potash mines (Annan et al., 1988).

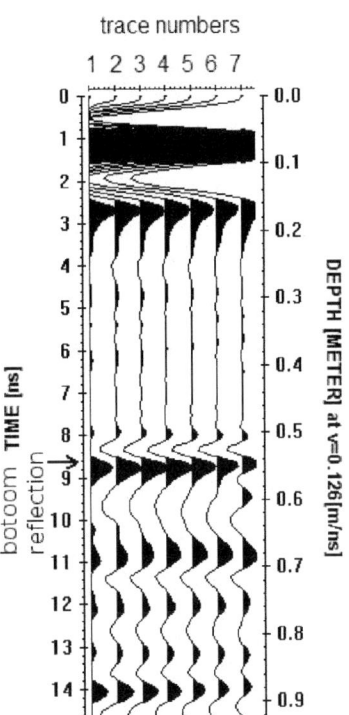

Figure 1.4. Radargram of the rock salt sample probing. The height of sample is 50.5 cm.

The bottom reflection is clearly identified at the times about 8 ns. Start time correction

and exponential gain function were applied to the data.

118

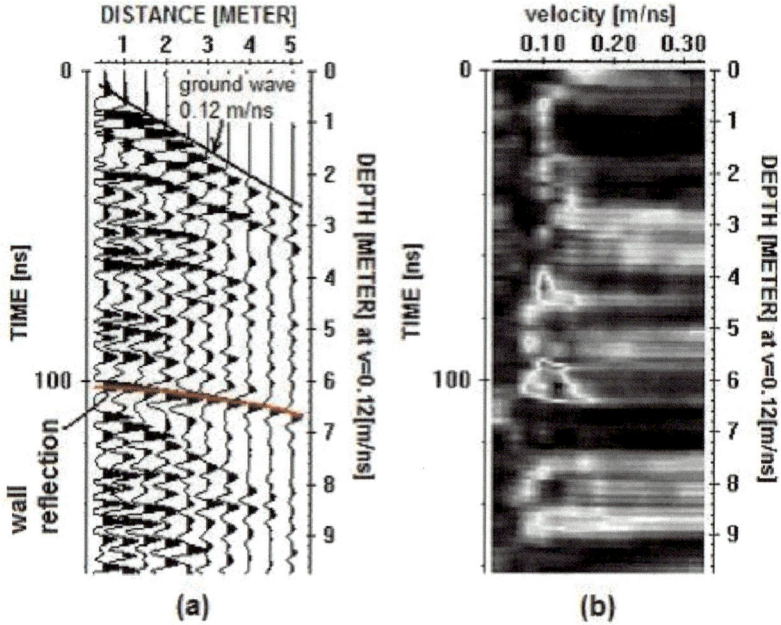

(a)  (b)

Figure 1.5. Semblance diagram (b) showing the velocity analysis of CMP data (a) acquired at the mine pillar of about 6 m thick using 250 MHz antennas. AGC (Automatic Gain Control) gain and start time corrections are applied to the data. Reflection from the opposite wall of mine pillar is reliably identified at the time about 100 ns. Other reflections are from the stress-relief cracks, fold limbs, and the edge of pillar. Semblance analysis demonstrates the fairly uniform velocities of 0.12 m/ns. The ground wave traveling at velocity of 0.12 m/ns is a primary event on the radargram. The direct airwave is suppressed with antennas shielding.

No studies of velocity variations in the Upper Kama potash mines were conducted. Mainly the bulk clay content controls velocity variations in the salt rocks of the Upper Kama potash deposit. An increase of velocity caused by fracturing (because of higher velocity in air) can be observed. It has been documented that stress-generated micro fracturing in concrete causes increasing the GPR velocities (Orlando et al., 2010). Despite the high plasticity of salt rocks, the intensive mine-induced forces are capable of producing significant amounts of micro fractures that can cause measurable changes of GPR velocities. The decrease of velocities in the moistened rock is expected, but salts of the Upper Kama deposit are usually dry and do not contain substantial volume of free water.

The insoluble content of the salt beds and the brine-saturated clay layers are factors that affect the radar penetration depth. No measurements of attenuation were made in the mine because of the complicated methodology required (Annan et al., 1988). However, attenuation coefficients of 1-2 db/m, which were reported in published literature (Annan et al., 1988), can be used in most cases for estimation of radio waves propagation characteristics in salts.

A summary of physical properties of the materials, which may be encountered in the Upper Kama potash mines, and the source of the data are given in Table 1.1.

Reflections are produced at the boundaries between materials having different electrical properties. Thin clay seams are the most prominent reflectors in the Upper Kama potash deposit. Because of the limited directivity of the antennas used in GPR systems, radargrams are often contaminated with out-of-plane reflections from fold limbs, adjacent tunnels and mine equipment. Also stress-relief fractures and folds produce numerous diffractions complicating the radargram interpretation.

Table 1.1. Measured electrical properties of materials that may be encountered in the Upper Kama potash mines.

| Materials | Dielectric constant | Velocity, m/ns | Source of data |
|---|---|---|---|
| Rock salt | 5.85 | 0.124 | Rock sample test |
| Rock salt | 5.67 | 0.126 | Rock sample test |
| Rock salt | 6.46 | 0.118 | WARR sounding |
| Sylvinite | 6.25 | 0.12 | CMP sounding |
| Carnallite | 5.33 | 0.13 | Rock sample test |
| Marker Clay | 25 | 0.06 | GPR data inversion |

**1.4.2. GPR Data Acquisition and Processing.** More than 5 km of continuous common offset GPR profiles were acquired in 2 field days. For testing the salts electrical properties, a number step-mode CMP and WARR measurements were conducted. The data were acquired using the commercial GPR system OKO (Logical Systems, 2005). The antennas with operating frequencies 400 and 150 MHz have been chosen for measurements after trial testing of the antennas of operating frequency of 400, 250, 150 and 50 MHz as providing with resolution and penetrating depth necessary for detection and delineation of the target objects.

Studies were conducted at two sites; at mining level AB of sylvinite zone to investigate the open fractures, and at 20 m below in conveyor drifts for mapping the stratigraphy and geological structures of the underlying rock salt strata. For continuous common offset profiling in the mine, the antennas were pulled along the floor or carried along the walls and roof. Several step mode CMP and WARR soundings were conducted to estimate the propagation velocities in salts. Typical profiling in the mine is shown in Figure 1.6.

Figure 1.6. GPR sounding of a mine pillar using 400 MHz antennas. The flat surface of excavations in potash mines allows pulling the antennas along the walls for profiling.

Reflexw (Windows OS) and OpendTect (Linux OS) software were used to process and interpret of 2-D and 3-D data respectively. A common processing flow was applied to both 2-D and 3-D data including: start time adjustment; "dewow" filtering to suppress the very low-frequency noise; 2-D filtering to remove the "ringing" and direct waves. An exponential or linear gain function was used to amplify the weak signals and preserve the amplitude information.

The variety of shape, size, depth and geological condition of the target structures required the object-specific procedures to be applied to GPR data after the common processing. The specific procedures used for data processing for each object of investigation are discussed in the next sections.

## 1.5. RESULTS AND DISCUSSION

A large volume of GPR data was acquired during the experimental field work. The most characteristic examples of GPR-mapped evaporite deformation are described in this Section.

**1.5.1. Site 1.** The principal objective of the Site 1 survey was to investigate the fracture zone exposed by workings at the AB mining level of South-East Panel 4 of the Solikamsk potash mine 3. The 400 MHz operating frequency antennas were chosen for fracture measurements providing both the necessary resolution, detectability and penetration depth.

The open fractures were expected to produce a prominent reflection because of the contrasting electric properties of rock and air. To estimate the feasibility of the GPR method to detect the millimeter opening fracture, finite-difference time-domain (FDTD) modeling was conducted. A model of a fracture with a 1 mm opening and a dip angle of $25°$ were chosen to replicate the parameters of fractures identified at the Site 1 study area. Results of modeling confirmed that the millimetric scale fracture could be reliably detected (but not resolved) using 400 MHz GPR antennas (Figure 1.7).

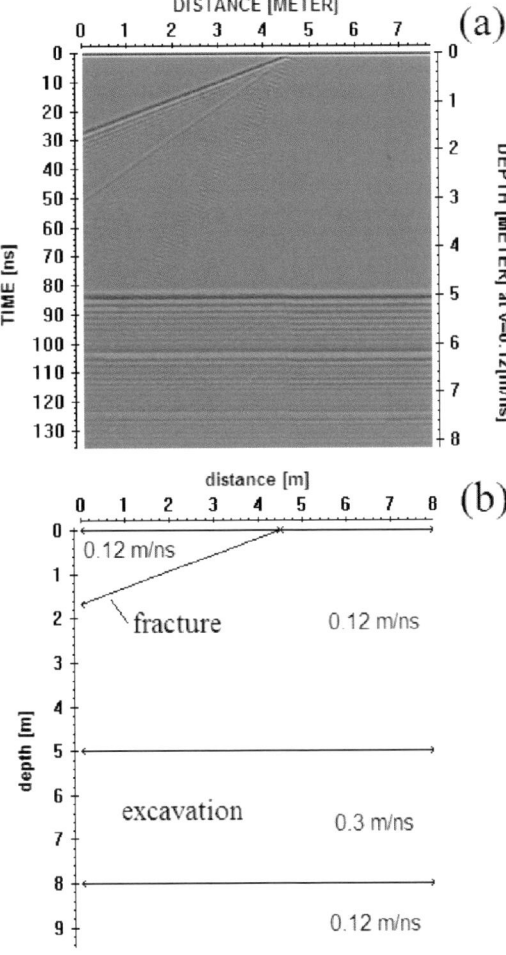

Figure 1.7. Georadar FDTD modeling data demonstrates the response of 1 mm opening

fracture in a 5 m thick mine pillar. (a) Georadar model of the fracture response for a

signal of frequency of 400 MHz. (b) Medium model used for calculation. Adjacent tunnel

is included to replicate the experimental environment. The reflection from the fracture

and reverberation in the tunnel are fairly easily identified. The fracture on the model is

detected but not resolved.

Seven profiles of 2-D continuous constant offset GPR data were collected along the walls in chambers 164, 166, 168, 173 of 200 m, and in the exploration drift (Figure 1.8). The trace spacing was 0.1 m.

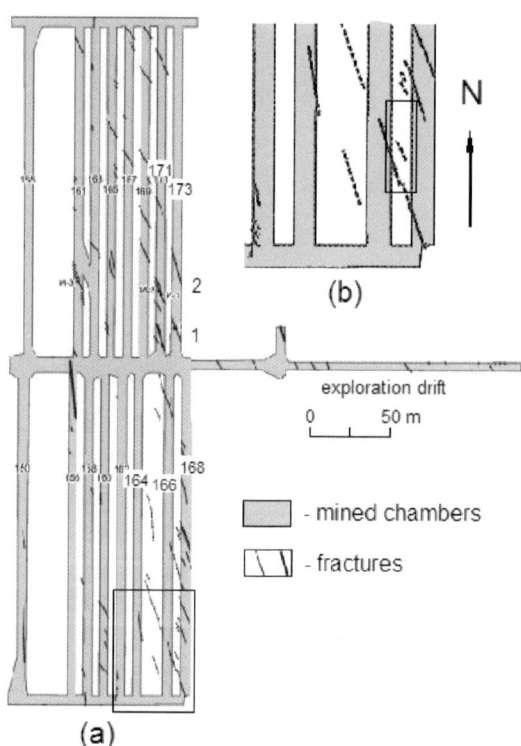

Figure 1.8. Layout of the study site on mining level AB of panel 4 (modified from Kudryashov et al., 2004). (a) Map of mine chambers (stops) and pillars shows the fractures observed on the surface of excavations and detected using georadar. (b) Location of the 3-D survey site. It also is marked on the general map by rectangle.

Planar events with a NW strike were clearly identified on the GPR profiling data (Figure 1.9). They coincide with fractures visually observed on the surface of mine excavations. A number of the planar reflectors not exposed on the surface were detected inside the rock mass. They have the signature and orientation similar to known features that supporting interpreting them as fractures.

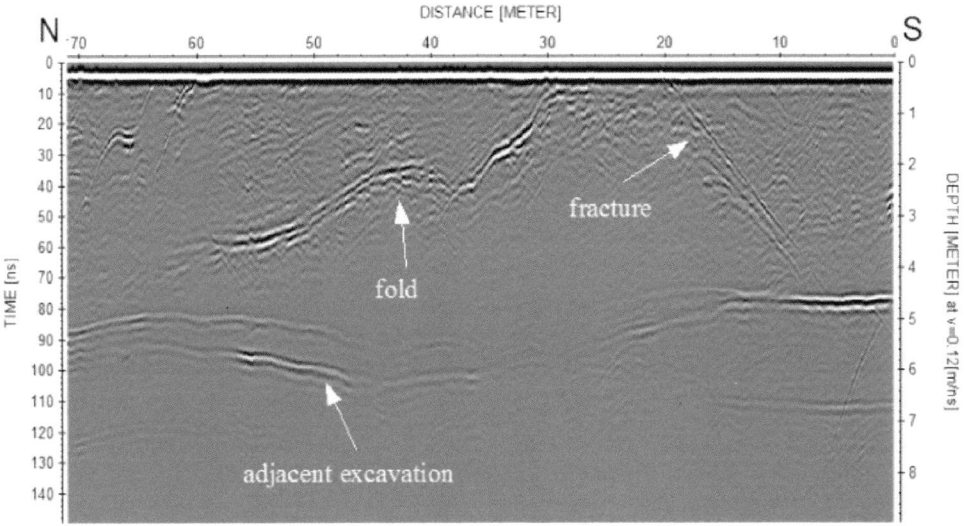

Figure 1.9. GPR profile acquired along the east wall of chamber 166 showing the common reflectors observed into the mine pillars. Start time correction and linear gain were applied to the data. Reflections from the fold limbs, fractures and walls of adjacent tunnel having different shape and orientation are clearly identified. The 400 MHz antennas were used for data acquisition. The start time correction and constant time gain are applied to the data. The profile was acquired from south to north.

Reflections from the fractures are locally superimposed by reflections from the fold limbs. Relatively uniform orientation and continuous signature allowed using of the FK filtering to improve the fracture mapping. An example of data acquired along the east wall of the Chamber 173 is shown in Figure 1.10. Reflections from the folded clay layers were successfully removed from the record after FK filtering. Two events corresponded to the visually verified fractures are marked on the map in Figure 1.8a and on the radargram.

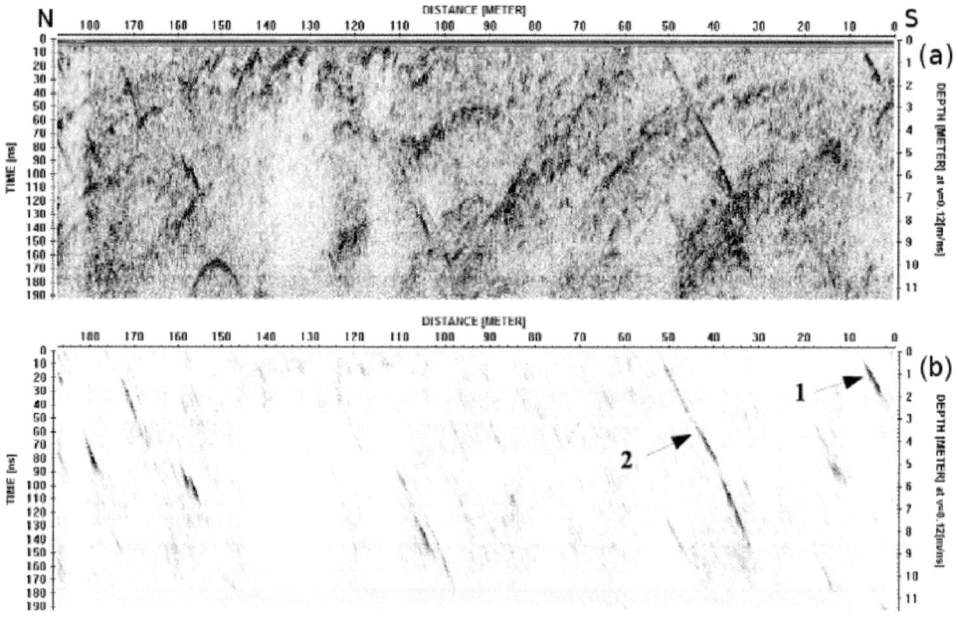

Figure 1.10. GPR profile at the east wall of chamber 173. (a) Data after start time and gain correction. (b) Data after FK filtering. Reflections from the folds are removed and fractures are fairly easy to recognize. The events marked by numbers 1 and 2 correspond to the fractures observed on the wall of excavation shown in Figure 1.8a.

The data collected on the roof and floor of the chambers showed that imaging of fractures of steep dip angles (near 90°) without significant vertical displacement, is rather problematic. Despite the 0.01 m trace spacing, only separate diffractions originating from irregularities on the surface and edges of fractures were observed (Figure 1.11).

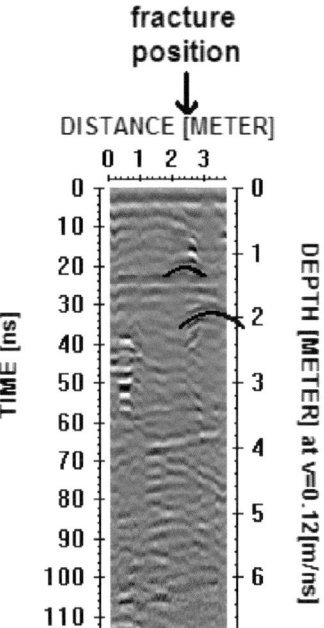

Figure 1.11. The GPR profile of an excavation roof across the visible fracture. The data were acquired using 400 MHz antennas. The trace spacing was 0.01 m. No reflections from the fracture plane are observed on georadar section. Scattering from irregularities on the fracture surface or at the fracture edges presumably generated the diffraction patterns indicated on the radargram by hyperbolas.

A small volume of 3-D data was collected across the visually observed fracture in the chamber 166 (Figure 1.8b). Using a 400 MHz antennas, seven horizontal parallel profiles of 8 m in length and spaced at 0.2 m interval were acquired across the fracture exposed on the east wall of mine pillar between the chambers 166 and 168 (Figure 1.12).

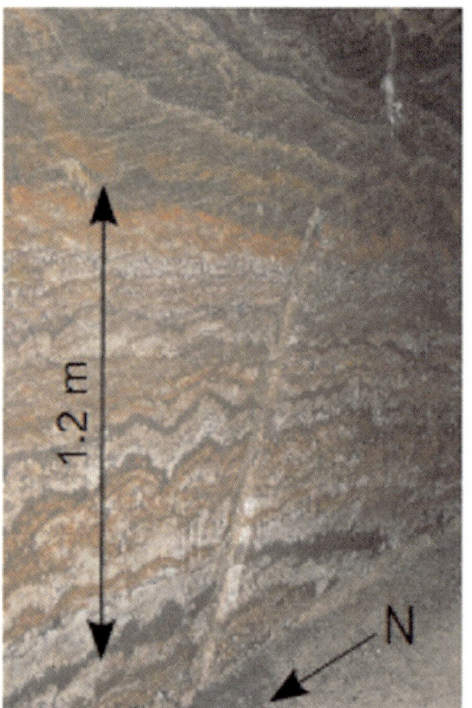

Figure 1.12. A fracture at the east wall of chamber 166 was used for 3-D experiment. The fracture of millimeter scale is gradually getting closed in upward direction.

Constant offset continuous profiling mode with 0.05 m trace spacing was used. The data were digitized at 0.585 ns time interval. The topography correction was not required because of flat surface of test site.

After common 2-D processing with Reflexw software (Sandmeier, 2003), the GPR data were converted to the SEG-Y format and loaded to the OpendTect seismic post-processing and interpretation system. A 3-D Kirchhoff migration algorithm of the Madagascar seismic data processing plugin available in the Linux version of the OpendTect was used to produce the 3-D image (Figure 1.13). A constant velocity 0.12 m/ns was used for migration. The seismic software does not interpret correctly the parameters of the GPR record because the time sampling rate and distance in the SEG-Y file header are written in form of integer number of microseconds and meters/feet. Therefore, the processing parameters were input according to the radar-to-seismic equivalence relationships: 1000 ns – 1 s, 100 MHz – 100 Hz, 0.1 m/ns – 1000m/s, and 1 m – 10 m (Grasmuek, 1996).

Using orientation, amplitude, and continuous signature of the known fracture as an interpretation criterion, the planar dipping reflector occurred between 2 and 4 m depth was interpreted to be a fracture. The amplitude of signal along this reflection varies suggesting a variation in fracture opening. Planar reflections parallel to the observation surface at a distance of about 5 m corresponds to the opposite wall of pillar. The time/depth slices and X, Y cuts were built, and the target events were picked throughout the data volume to create a numerical spatial model of subsurface structures (Figure 1.14).

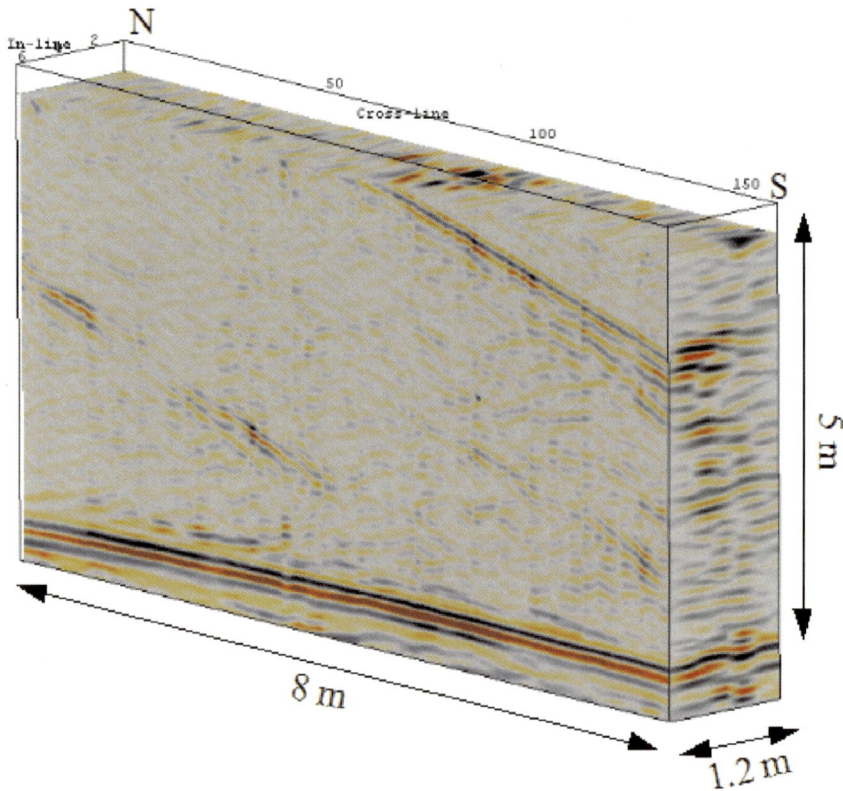

Figure 1.13. The GPR data volume obtained after 3-D Kirchhoff migration. Reflection at the depth interval from 2 m to 4 m at the northern part of data cube was interpreted as a fracture. Note, that orientation of the data cube is conventional for 3-D data visualization layout with a vertical depth coordinate axis. In the case of profiling along the tunnel walls the depth axis is in the horizontal plane.

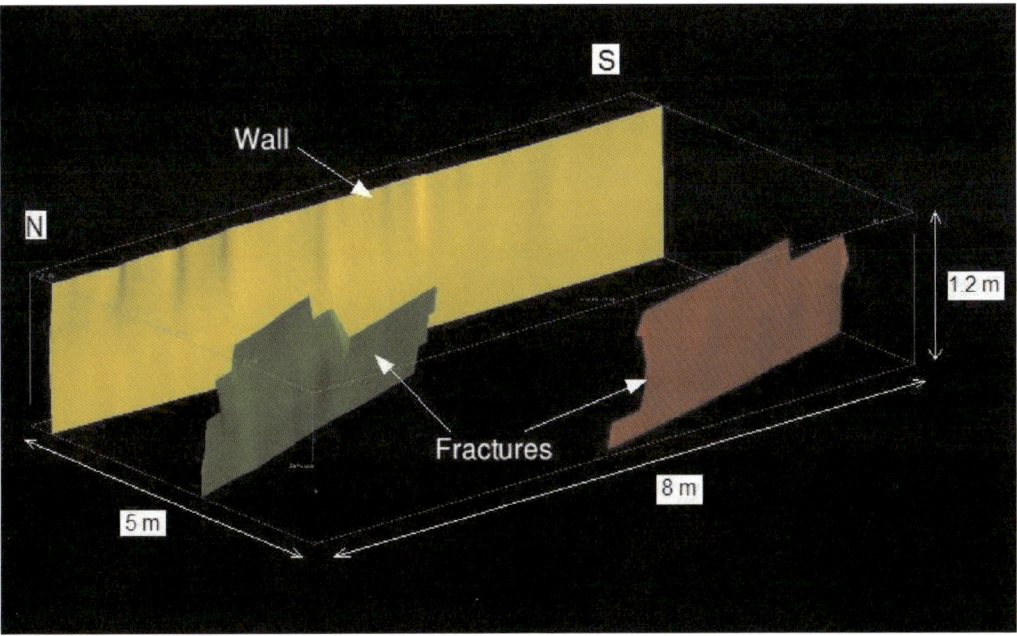

Figure 1.14. Spatial model of reflectors within the mine pillar derived from the 3-D GPR image. Orientation of the model corresponds to structures actual position in the mine.

**1.5.2. Site 2.** Measurements in the underlying rock salt were conducted in order to map the stratigraphy and deformational structures beneath the mine workings. The common offset profiling data were acquired in the conveyor drift, brine collector drift, and in the Red II entry tunnel of the Block 1 of South-East panel 4.

The 50 MHz and 150 MHz antennas were used for profiling along the floor of excavations. Both antennas demonstrated the same penetrating range because the thick Marker Clay layer at approximately 20 m below the conveyer drift controls the penetration depth. Therefore, the 150 MHz antennas, which provide higher resolution, were chosen for measurements.

Analysis of acquired data showed that GPR method is able to accurately delineate geological structures in uniform salt rocks. Example GPR data from the entry tunnel at the western end of Block 1 (Figure 1.15) images folds of different ranges and shapes that commonly occurs in salt rocks (Figure 1.16a). Reflections are generated dominantly by the thin centimeter scale clay layers. The main reflector in the underlying rock salt is a Marker Clay layer (indicated as MC) of thickness ranging approximately from 1 to 2 m.

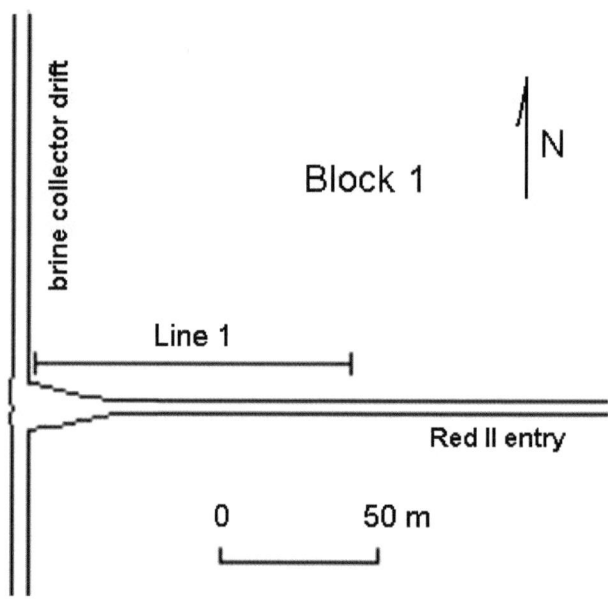

Figure 1.15. Layout of measurements in the Red II entry tunnel.

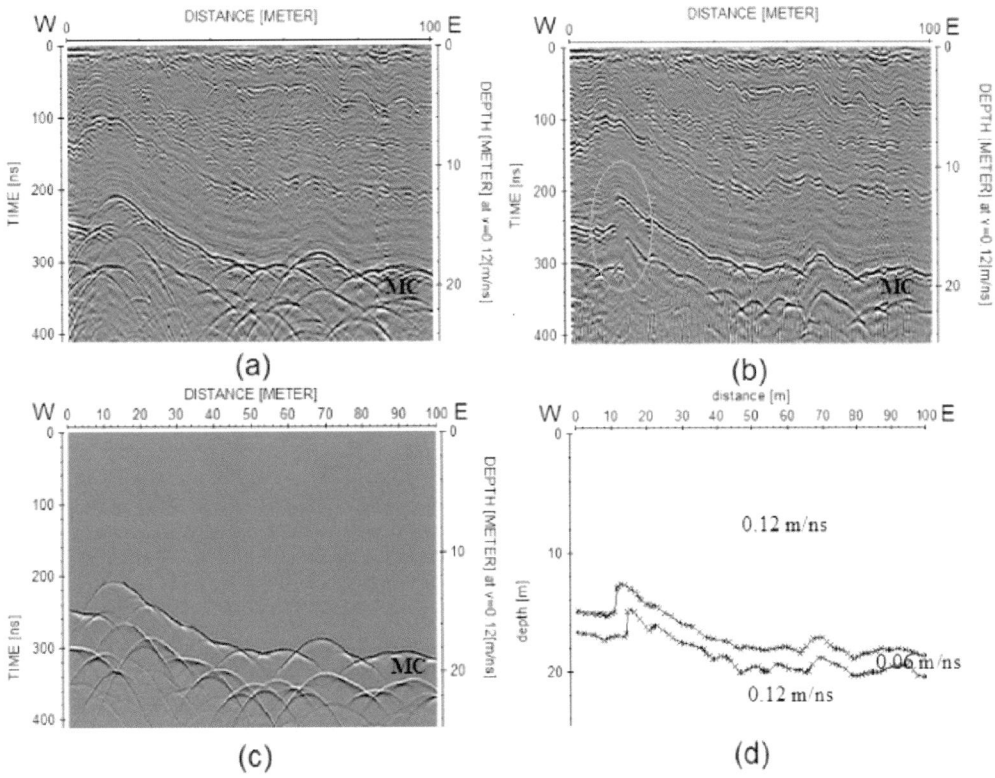

Figure 1.16. Example of common offset data acquired in the underlying rock salt with 150 MHz antennas. (a) Raw data after start time correction and gain decay compensation. (b) Georadar section after FD-migration. (c) Resulting section of FDTD modeling. (d) Model of Marker Clay layer retrieved from the migrated profile data. The assumed fault structure is indicated by circle. Note, the GPR profiles are vertically exaggerated; the folds have more gentle geometry than this visual presentation might suggest.

The data were significantly contaminated with diffraction hyperbolas in the folded layers that make the interpretation of the target events difficult (Figure 1.16a). Diffraction concentrations increase with depth because of the lower frequency content of deeper reflections caused by the signal dispersion. GPR wave's dispersion, caused by frequency-dependent attenuation, is a common effect of signal propagation in geological material. It results in the decrease of central frequency of the signal with depth (Irvine and Knight, 2000). The FD migration algorithm was used to remove the diffractions from the radargram and restore the true geometry of reflection boundaries. The 2-D velocity model needed for migration was obtained through interactive hyperbola adaptation.

Resulting data provided a detailed image of plastic deformation structures within the salt rock (Figure 1.16b). The asymmetric folds on the section have more deep western limbs. It is consistent to east-to-west direction of principal regional stress. Displacement (indicated by a circle) of about 2.5 m was revealed at the Marker Clay. This structure was interpreted as fault or fault propagation fold. The fault is gradually died out within the overlying strata.

Modeling was used to verify and better understand the interpretation. The finite-difference (FD) modeling program of Reflexw package was employed to simulate the reflections from the Marker Clay layer. The initial model of Marker Clay layer topography was retrieved from migrated data (Figure 1.16d).

There is little information about variations in the thickness of the Marker Clay and no information about its dielectric properties. Mine personnel provided the estimated thickness of the Marker Clay layer of approximately 1.6 m. A simple iterative inversion technique was used to determine the propagation velocity of the clay layer. The velocity in the layer was iteratively changed until a time thickness of the layer at the reference points on the model section and unmigrated GPR profile were comparable.

Through data inversion the propagation velocity in Marker Clay layer of 0.06 m/ns was determined. This value is lower than expected because the clays of salt deposits are usually salt saturated and dry relative to those on the surface. The electrical properties of thin clay seams given in Annan et al., 1988 also are very close to the properties of salts. The properties of thick clay layers have not been studied before and can differ from those of thin seams. Therefore, the defined value can be accepted as approximate and further study of clay layers properties must be undertaken to make the interpretation more reliable.

The model of reflections from the Marker Clay layer calculated using defined parameters is presented in Figure 1.16c. It fairly satisfies the real data.

**1.5.3. Penetration Depth.** The penetration depth of the GPR signal in common geological materials is controlled by conduction losses, water relaxation, electrochemical processes on the surface of clay particles, scattering on the heterogeneities in the medium, and central frequency of the antennas employed. The investigation depth increases for low frequency signals. In lossless pure rock salt, electromagnetic waves are capable of penetrating hundreds of meters. In the Upper Kama potash deposit, the main factor affecting the electromagnetic signal penetration is the presence of clay layers. Both, conduction losses and scattering at the interface salt-clay occur. For example, the Marker Clay layer entirely attenuates GPR signal and is a major factor governing the penetration depth in underlying rock salt. Locally, the fracturing, adjacent openings and brine-saturated salt rocks significantly limit the penetration. The penetration depths of up to 10 m and up to 20 m were achieved for 400 MHz and 150 MHz antennas respectively.

Recent research (Orlando et al., 2010) demonstrated the increase of the GPR velocities and attenuation caused by stress-generated micro fracturing in the concrete pier. This effect is related to increase of the air-filled space into the material. Because the GPR velocity in air is higher than in material, an increase in the average velocity of the fractured media occurs. Fracturing increases the attenuation of the GPR signal and can be an additional factor decreasing the penetration depth.

The supporting pillars are the elements of mine structure that have been subjected to considerable compressional forces. This results in creep, micro-fracturing and, finally, development of open fractures. Monitoring stress changes in the mine pillars is one of the most important problems for estimating the mine stability. It is possible that the velocity determination methodology developed in this study might be used to monitor fracture development in the mine pillars in the Upper Kama potash mines.

**1.5.4. Resolution and Detectability.** Resolution and detectability are usually used for characterizing the capability of the GPR method to locate and determine parameters of the target objects. The vertical resolution of GPR data defines a minimum distance between two reflectors necessary for them to be distinguished on a georadar record. Theoretically, the minimum resolvable thickness of a layer is one-quarter of wavelength (Sheriff and Geldart, 1995). Therefore, resolution increases with increasing frequency. The vertical resolution in salt rock having a velocity 0.12 m/ns achieves 0.07 m at operating frequency of 400 MHz, 0.2 m using 150 MHz and 0.6 m using 50 MHz antennas. Using a 400 MHz antenna, the width of the air-filled fracture might be defined if the opening is more than 0.19 m and the fractures of millimeter scale cannot be resolved. However, in this and some other studies, fractures, considerably thinner than the GPR wavelength employed, are readily detectable (Lane et al., 2000; Annan, 2003; Kovin, 2010).

The investigated fractures at the test site extended to depths of several meters so ultra-high frequency antenna, which would have provided maximum resolution, were not utilized. Because intermediate frequency 400 MHz antennas were used, the detectability of fractures (rather than resolution) was the priority of this experimental work. Vertical detectability (simply detectability is used herein) is defined as the capability to detect (visually identify) thin subsurface layers that cannot be resolved using signal of given frequency. Detectability (sensitivity) of thin fractures depends upon many factors (central frequency and shape of signal, fracture opening, variation of fracture geometry, roughness of the walls, orientation relative to the antenna and observation line, homogeneity of surrounding media, filling material, parameters of recording, parameters of the processing, etc.). The millimeter scale fractures were successfully detected by Lane et al. (2000) and Laurence et al. (2003) but in other cases the fractures were not detected (Apel and Dezelic, 2005b).

The total reflected signal from a thin layer is formed by the constructive and destructive superposition of the primary reflections from the walls and the multiples bouncing between interfaces with a phase shift depending on the thickness and velocity in the layer material (Annan, 2001). The amplitude of reflected signal is proportional to the difference in the electromagnetic properties at the boundary of two materials. In case of salt-air and air-salt interfaces, related to air-filled fracture in salt rock, two primary partial reflections have opposed phases and may interfere destructively if the opening of the fracture is smaller than wavelength of electromagnetic pulse. A decrease in the fracture opening leads to a decrease in the amplitude of the total reflection because of smaller time delay between the reverse polarity almost superimposed reflections from the salt/air and air/salt interfaces. The relative degradation/diminishment of total reflection from thin fracture depends on the reflection coefficient of the first interface (sat/air) because the relative amplitude of the destructive reflection from opposed interface (air/salt) of the fracture depends, in large part, on the portion of energy passed through the interface. Increasing of reflection coefficient of the first interface increases the amplitude of total reflection. Moisture, condensed on the walls of fractures, increases the amplitude of total thin fracture reflection because it increases the reflection coefficient of the first interface thereby minimizing the amplitude of the destructive reflection from second interface. This effect significantly improves the detectability of thin fractures (Apel and Dezelic, 2005b; Leucci et al., 2007).

Scattering of electromagnetic waves on the irregularities in shape and rough surface of fracture walls can significantly improve the detectability of thin fractures because it increases the partial energy reflected from the first boundary. For identification of weak reflections from thin or small targets, the low noise level on the record is very important. The salt rocks are relatively uniform and, hence, are characterized by low level of structural noise that allows recording the weak signals and enhance the detectability of thin fractures. The signal reflected from thin layer is close to the time derivative of the incident wave (Annan, 2001). It increases the frequency of reflected signal (see Figure 1.7) and suggests using broadband antennas to effectively receive the reflection from thin fracture.

It is difficult to predict the detectability of fracture because the many factors govern the behavior of GPR signal within natural rock mass. But simplified estimates can be made. For example, the scheme of partial reflection energy for air-filled fracture (dielectric permittivity = 1) within salt rock (dielectric permittivity = 5) is shown in Figure 1.17. The reflection coefficient calculated using equation (55) allows the estimation of the approximate energy and amplitude of total reflection that is produced by superposition of partial reflections from salt-air and air-salt boundaries of fracture. The reflection coefficient of first interface salt/air is ~ 0.38. The "effective" reflection coefficient from the air/salt interface is 0.15. The destructive reflection from the second interface (air/salt) will therefore be markedly lowered because of this energy loss due to partial reflection. This simplistic approach demonstrates why thin air-filled fractures within salt rock are detectable (but not resolved). Note that the shape of total reflected signal varies with relationship of wavelength and fracture opening and can be defined using modeling algorithms. An example of the finite-difference model of air-filled fracture of 1 mm opening within salt rock for signal frequency of 400 MHz is shown in Figure 1.7.

There is a trade-off between the resolution and penetration depth because the higher frequencies are attenuated more rapidly with distance. Consequently, the optimal choice of antennas must reflect both the depth and size of the object. The antenna with operating frequencies 400 and 150 MHz have been chosen for further investigation after trial testing of the antennas of operating frequency of 400, 250, 150 and 50 MHz. Both antennas were shielded that prevented the data from being contaminated by reverberation within the tunnel and reflections from mine equipment.

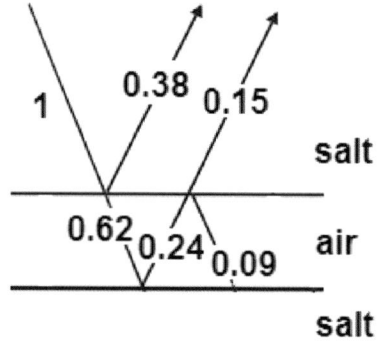

Figure 1.17. Scheme of partial reflected energy from air-filled fracture within salt rock. Numbers in the picture show the energy related to each partial reflection outside and inside the facture. The energy of incident signal is assumed to be 1. The absolute value of reflection coefficients of interfaces salt/air and air/salt is 0.38. Reflection from the interface air/salt has reverse polarity.

## 1.6. CONCLUSIONS

Experimental data were acquired in the underground mine of the Upper Kama potash deposit in order to assess the capability of GPR method for mapping the brittle and ductile deformational structures in salt rocks.

Although the GPR method has been extensively used in German and Canadian salt mines, studies have shown that the GPR acquisition, data processing, and interpretation methodologies required an adaptation to local geological and mining environment.

The results of study demonstrate that GPR method is capable to reliably detect and delineate the open fractures of up to $25°$ deep angles, faults and folds within evaporite strata of the Upper Kama potash deposit. Unfortunately, the imaging of subvertical fractures appeared problematic.

The fracture detection was significantly improved using FK filtering. Because of strong folding the GPR records are contaminated with diffractions and migration of the data is required to make the radargrams interpretable. Usage of 3-D GPR imaging technique allowed accurate delineating the fractures.

Usage of modeling improved the understanding the deformational structures occurred within the salt rocks of the Upper Kama potash deposit.

## 1.7. ACKNOWLEDGEMENTS

The author thanks the Department of Geological Sciences & Engineering at the Missouri University of Science and Technology for financial support of this research project. I especially appreciate the Geological Survey of JSC "Silvinit" for the kindly permission and assistance to GPR survey in their potash mine. I am also very grateful to the company "Orgtechstroy" for allowing GPR equipment and Sergey Baluyev for assistance in conducting the field work.

# 5. CONCLUSIONS

Experimental GPR data were acquired in the Solikamsk potash mine 3 of the Upper Kama potash deposit to study the capability of method for mapping geological structures in salt rocks and development methodology of its usage in local geological and mining environment.

Analysis of the acquired data has shown that the usage of GPR method in potash mines of the Upper Kama deposit have some peculiarities that differ from those in the German and Canadian potash mines that requires an adaptation of the acquisition, data processing, and interpretation methodologies to local geological and mining environment. For example, the interpretation of reflections from geological boundaries is significantly complicated because of multiple diffraction hyperbolas produced by the widespread multi-scale folding. Migration should be performed to make the data interpretable and get a true geometry of the folded beds.

The object-oriented processing schemes were developed to effectively detect and delineate the subsurface structures in geological conditions of the Upper Kama potash deposit.

Usage of FK filtering significantly improved the fractures detection. The spatial model of fractures was created using 3-D imaging technique.

Non-destructive, high-resolution, portable and cost-effective method, GPR appears well suited for usage in the salt mine environment.

The results of data analysis show that GPR method may be of great utility for detection of the small-scale fractures, folds of various range, and fault structures in the salt rocks at the depth of up to 20 m. The 3-D imaging technique has proven to be an effective tool for studying the geometry of fractures.

This study has the following results:

- the electrical properties of typical evaporite formation members were determined;

- the site-specific data acquisition technique and object-oriented data processing schemes adapted to the local geological and geotechnical environment are developed;

- the methodology of 2-D and 3-D GPR data interpretation is worked out.

The further research must be aimed to study of the electromagnetic properties of evaporite rocks to improve the interpretation of the GPR data obtained in potash mines. Because the monitoring of the stress state in the supporting pillars is very important for mine safety, the studies of relation between parameters of GPR signal and stress will be recommended.

# BIBLIOGRAPHY

Anderson, N.L., Brown, R.J., 1992. Dissolution and deformation of rock salt, Stettler area, Southeastern Alberta. Canadian Journal of Exploration Geophysics 28 (2), 128-136.

Andrejchuk, V., 2002. Collapse above the world's largest potash mine (Ural, Russia). Int. J. Speleol. Vol. 31 (1/4), 137-158.

Annan, A.P., Davis, J.L., Gendzwill, D., 1988. Radar sounding in potash mines, Saskatchewan, Canada. Geophysics 53 (12), 1556-1564.

Annan, A.P., 2001, Ground Penetrating Radar. Workshop Notes. Sensors & Software, Inc., Mississauga, Canada.

Annan, A.P., 2002. GPR – History, Trends, and Future Developments. Subsurface Sensing Technologies and Applications 3 (4), 253 – 270.

Annan, A. P., 2003. Ground penetrating radar: Principles, procedures and applications. Mississauga, Ontario, Sensors & Software, Inc.

Annan, A.P., 2009. Electromagnetic principles of Ground Penetrating Radar. *in* Ground Penetrating Radar: Theory and Applications. Ed. H.M. Jol. Elsevier B.V., The Netherlands, 1-40.

Apel, D.B., Dezelic, V., 2005a. Using ground penetrating radar in analyzing structural composition of roofs in tunnels. SME Journal 60 (7), 56-60.

Apel, D.B., Dezelic, V., 2005b. Evaluation of high frequency ground penetrating radar (GPR) in mapping strata of dolomite and limestone rocks for ripping technique. International Journal of Surface Mining, Reclamation and Environment 19 (4), 260-275.

Baker, G. S., Jordan, T. and Pardy, J., 2007. Introduction to Ground Penetrating Radar (GPR). *in* Stratigraphic Analyses Using GPR. Ed. Gregory Baker and Harry Jol. Boulder, Colorado: Geological Society of America, 1-18.

Belkin, V.V., 2010. Environmental transformations within the Upper Kama evaporites basin. Doctor of Science Thesis, The Ural State Mining University.

Brady, B.H.G., Brown, E.T., 2007. Rock mechanics: For underground mining. Springer, The Netherlands.

Busby, J.P., Merritt, J.W., 1999. Quaternary deformation mapping with ground penetrating radar. Journal of Applied Geophysics 41, 75-91.

Chaykovskiy, I.I., Gorbunov, S.G., Korochkina, O.F., Andreyko, L.V., 2009. Solikamsk depression. *in* Geologic Heritage of Permskiy kray, http://perm-kray.ru/pam007-2.htm , July 2010.

Chouteau, M., Phillips, G., Prugger, A., 1997. Mapping and monitoring softrock mining. Proceedings of Exploration 97: Fourth Decennial International Conference on Mineral Exploration, Toronto, Canada, September 9-12, 927-939.

Christie, M., Tsoflias, G.P., Stockli, G.F., Black, R., 2009. Assessing fault displacement and off-fault deformation in an extensional tectonic setting using 3-D ground-penetrating radar imaging. Journal of Applied Geophysics 68, 9-16.

Cook, J.C., 1969. Electromagnetic exploration within Salt Domes. Proceedings of The Third Symposium on Salt, Cleveland, April 21-24, 386-390.

Cook, J.C., 1975. Radar transparencies of mine and tunnel rocks. Geophysics 40 (5), 865-885.

Coon, J.B., Fowler, J.C., and Schafers, C.J., 1981. Experimental uses of short pulse radar in coal seams. Geophysics 46 (8), 1163-1168.

Daniels, J.J., 1996. Surface Penetrating Radar. The Institution of Electrical Engineers, London.

Daniels, D.J., 2009. Antennas. *in* Ground Penetrating Radar: Theory and Applications. Ed. H.M. Jol. Elsevier B.V., The Netherlands, 99-139.

Davis, J.L., Annan, A.P., 1989. Ground-penetrating radar for high-resolution mapping of soil and rock stratigraphy. Geophysical Prospecting 37, 531-551.

Eso, R.A., Oldenburg, D.W., Maxwell, M., 2006. Application of 3-D electrical resistivity imaging in an underground potash mine. SEG Expanded Abstracts 25, 629-632.

Friberg, M., Juhlin, C., Beckholmen, M., Petrov, G.A., Green, A.G., 2002. Palaeozoic tectonic evolution of the Middle Urals in the light of the ESRU seismic experiments. Journal of the Geological Society 159, 295–306.

Garrett, D.E., 1995. Potash: Deposits, Processing, Properties and Uses. Chapman & Hall, London.

Gendzwill, D., 1969. Underground applications of seismic measurements in a Saskatchewan potash mine. Geophysics 34 (60), 906-915.

Gendzwill, D.J., Stead, D., 1992. Rock mass characterization around Saskatchewan potash mine opening using geophysical techniques: a review. Canadian Geotechnical Journal 29 (4), 666-674.

Geography, 2009. Berezniki collapse. Geography, 5, (in Russian) <http://geo.1september.ru/view_article.php?id=200900505> , July 2010.

Glebov, S.V., 2006. Substitution of the rational integrated geophysical investigation of water protective strata at the soluble ore deposits: Upper Kama Potash Deposit. Candidate of Science Thesis, Mining Institute UB RAS, Perm (in Russian)

Gottsche, F. M., 1997. Identification of cavities by extraction of characteristic parameters form ground probing radar reflection data. PhD thesis, Kiel University.

Gowan, S.W., Trader, S.M., 2000. Mine Failure Associated with a Pressurized Brine Horizon: Retsof Salt Mine, Western New York. Environmental & Engineering Geoscience VI (1), 57-70.

Grandjean, G., Gourry, J.C., 1996. GPR Data processing for 3D fracture mapping in a marble quarry (Thassos, Greece). Journal of Applied Geophysics 36, 19-30.

Grasmueck, M., 1996. 3D ground penetrating radar applied to fracture imaging in gneiss. Geophysics 61 (4), 1050-1064.

Grasmueck, M., Weger, R., Horstmeyer, H., 2004. Three-dimensional ground-penetrating radar imaging of sedimentary structures, fractures, and archeological features at submeter resolution. Geology 32 (11), 933-936.

Gregoire, C., Halleux, L., 2002. Characterisation of fractures by GPR in a mining environment. First Break 20 (7), 467-471.

Griffiths, D. J., 1999. *Introduction to electrodynamics.* (Third Edition ed.). Prentice Hall, 559 pp.

Gross, R., Green, A., Holliger, K., Horstmeyer, H., Baldwin, J., 2002. Shallow geometry and displacements on the San Andreas Fault near Point Arena based on trenching and 3-D georadar surveying. Geophysical Research Letters 29 (20), 34-1-34-4.

Gross, R., Green, A., Horstmeyer, H., Holliger, K., Baldwin, J., 2003. 3-D georadar images of an active fault: efficient data acquisition, processing and interpretation strategies. Subsurface Sensing Technologies and Applications, 4 (1), 19-40.

Gross, R., Green, A., Horstmeyer, H., 2004. Location and geometry of the Wellington Fault (New Zealand) defined by detailed three-dimensional georadar data. Journal of Geophysical Research 109 (B05401), 1-14.

Haeni, F.P., Halleux, L., Johnson, C.D., Lane, J.W., 2002. Detection and mapping of fractures and cavities using borehole radar. *in* Fractured Rock 2002, Denver, Colorado, March 13-15, 4p.

Hayt, W. H., 1989. *Engineering electromagnetics.* McGraw Hill, Inc., USA.

Holser, W.T., Brown, R.J.S., Roberts, F.A., Fredrikkson, O.A., Unterberger, R.R., 1972. Radar logging of a salt dome. Geophysics 37(5), 889-906.

Irvine, J., Knight, R., 2000. Estimation and correction of wavelet dispersion in GPR data. Proceedings of 8th International Conference on Ground Penetrating Radar, pp. 561-567.

Jeremic, M.L., 1994. Rock mechanics in salt mining. A. A. Balkema, Rotterdam.

Kopnin, V.I., 1995. Verkhnekamskoye deposit of potassium, potassium-magnesium, rock salt, and natural brines. Gorniy Zhurnal (Mining Journal) 6, 10-43 (in Russian).

Kovin, O.N., 2000. Some results of acoustic reflection testing in Russian potash mines. Journal of Environmental and Engineering Geophysics 5 (1), 39-45.

Kovin, O.N., 2002. Investigation of the salt rocks deformations with georadar sounding technique. Proc. Annual Sci. Workshop, Mining Institute UB RAS, Perm, Russia, 30-32 (in Russian).

Kovin, O.N., Mironov, C.A., Kvitkin, S.Y., Kuznetsov, N.V., 2002. Case history: application of georadar for solving mining problems in potash mines of JSC "Uralkali". Proceedings of International Conference "Georadar 2002", Moscow State University, Moscow, January 28 – February 1, 22-23 (in Russian).

Kovin, O.N., 2010. Fractures imaging in the Upper Kama potash mine using3-D GPR data. Proceedings of the Symposium on the application of geophysics to engineering and environmental problems (SAGEEP 2010), Keystone, Colorado, April 11-15, cd-rom.

Konstantinova, S.A., Chernopazov, S.A., Gulyaev, A.A., 2001. Estimate of initial stresses in rock mass of the Upper Kama region based on block hierarchical model. Journal of Mining Science 37 (5), 447-454.

Konstantinova, S.A., Khronusov, V.V., 1999. Rock pressure manifestation around underground workings in potassium mines in the case of a nonhydrostatic initial stressed state of the rock mass. Journal of Mining Science 35 (2), 126-134.

Kudryashov, A.I., 2001. Upper Kama (Verkhnekamskoye) salt deposite. Mining Institute of Ural Branch of the Russian Academy of Sciences, Perm (in Russian).

Kudryashov, A.I., Vasyukov, V.E., Von-der-Flaass, G.S., Ikonnikov, E.A., Gershanok, V.A., Gershanok, L.A., Glebov, S.V., 2004. Rupture tectonics of the Upper Kama (Verkhnekamskoye) salt deposite. Mining Institute of Ural Branch of the Russian Academy of Sciences & Perm State University, Perm (in Russian).

Lane, J.W., Buursink, M.L., Haeni, F.P., Versteeg, R.G., 2000. Evaluation of ground-penetrating radar to detect free-phase hydrocarbons in fractured rocks – results of numerical modeling and physical experiments. Ground Water 38, 929-938.

Lajtai, E.Z., Carter, B.J., Duncan, E.J.S., 1994. En echelon crack-arrays in potash salt rock. Rock Mechanics and Rock Engineering 27 (2), 89-111.

Laurence, S., Balayssac, J.P., Rhazi, J., Klysz, G., Arliguie, G., 2003. Non destructive evaluation of concrete moisture by GPR technique: experimental study and direct modeling. Proceedings of International Simposium (NDT-CE 2003) cd-rom.

Leucci, G., Persico, R., Soldovieri, F., 2007. Detection of fractures from GPR data: the case history of the Cathedral of Otranto. Journal of Geophysics and Engineering 4, 452-461.

Liu, L., Li, Y., 2001. Identification of liquefaction and deformation features using ground penetrating radar in the New Madrid seismic zone, USA. Journal of Applied Geophysics 47, 199-215.

Lozovsky, V.R., Minikh, M.G., Grunt, T.A., Kukhtinov, D.A., Ponomarenko, A.G., and Sukacheva, I.D., 2009. The Ufimian Stage of the East European Scale: Status, Validity, and Correlation Potential. Stratigraphy and Geological Correlation 17 (6), 602-614.

Malovichko, A., Shulakov, D., Dyaguilev, R., Sabirov, R., Ahmetov, B., 2001. Comprehensive monitoring of the large mine-collapse at the Upper Kama Potash Deposit in Western Ural. Rockbursts and Seismicity in Mines – RaSiM5, South African Institute of Mining and Metallurgy, 309-312.

Malovichko, A.A., Sabirov, R.H., Ahmetov, B.Sh., 2005. Ten years of seismic monitoring in mines of the Verkhnekamskoye potash deposit. *in* Controlling Seismic Risk, Proc. of the Sixth International Symposium on Rockbursts and Seismicity in Mines (RaSiM6), 9-11 March, Australia, 367-372.

Martinez, J., Johnson, K., Neal, J., 1998. Sinkholes in evaporite rocks. American Scientist 86 (1), 38-51.

Maybee, G.W., Maloney, S., Kaiser, P., Triltzsch, G., Braun, H., Prugger, A., 2004. Stepped Frequency GPR Field Trials for Determining the Ideal Frequency Bandwidth for Use in Potash Mines. Potash – CIM, Edmonton, May 09-12, cd-rom.

Neal, J.T., Myers, R.E., 1995. Origin, diagnistics, and mitigation of a salt dissolution sinkhole at the US Strategic Petroleum Reserve storage site, Weeks Island, Louisiana. *in* Land Subsidence, Proceedings of the Fifth International Symposium on Land Subsidence, The Hague, October, 184-195.

Nesterov, M.P., Marakov, V.Y., 1995. Technogenic potash seams and their mining prospects. *in* Mine Planning and Equipment Selection. Ed. Singhal et al., Balkema, Rotterdam, 183-185.

Olhoeft, G.R., 1986. Electrical properties from $10^{-3}$ to $10^{9}$ Hz - physics and chemistry. *in* Physics and Chemistry of Porous Media II, American Institute of Physics Conf. Proc. 154, Ridgefield, CT, eds. J.R. Banavar, J. Koplik, and K.W. Winkler, NY, AIP, 281-298.

Orlando, L., 2003. Semiquantitative evaluation of massive rock quality using ground penetrating radar. Journal of Applied Geophysics 52, 1-9.

Orlando, L., Pezone, A., Colucci, A., 2010. Modeling and testing of high frequency GPR data for evaluation of structural deformation. NDT&E International 43, 216-230.

Polyanina, G.D., 1995. Development of underground mining technology for the potassium and potassium-magnesium salts of the Verkhnekamskoye deposit. Gorniy Zhurnal (Mining Journal) 6 (in Russian).

Porsani, J.L., Sauck, W.A., Junior, A.O.S., 2006. GPR for mapping fractures and as a guide for the extraction of ornamental granite from a quarry: A case study from southern Brazil. Journal of Applied Geophysics 58, 177-187.

Prugger, F.F., 1980. The flooding of the Cominco Potash Mine and its rehabilitation. Proc. Fifth Int. Simposium on Salt, Northern Ohio Geological Survey, 333-340.

Prugger, F.F., Prugger, A.F., 1991. Water problems in Saskatchewan potash mining. CIM Bulletin 84 (945), 58-66.

Rodgers, J., 1990. Fold-and-thrust belts in sedimentary rocks. Part 1: Typical examples. American Journal of Science, 290, 321-359.

Sandmeier, K.J., 2003. Reflexw version 3.0 (User's guide). Karlsruhe, Germany <www.sandmeier-geo.de> .

Schreiber, B.C., Helman, M.L., 2005. Criteria for distinguishing primary evaporite features from deformation features in sulfate evaporites. Journal of Sedimentary Research 75, 525-533.

Seol, S.J., Kim, J.H., Song, Y., Chung, S.H., 2001. Finding the strike direction of fractures using GPR. Geophysical Prospecting 49, 300-308.

Sheriff, R.E. and Geldart, L.P., 1995. Exploration Seismology, Cambridge University Press, 592 pp.

Stewart, R.D., Unterberger, R.R., 1976. Seeing through rock salt with radar. Geophysics 41 (1), 123-132.

Thoma, H., Lindner, U., Klippel, O., Schicht, T., 2003. Geophysical methods as one way to detect and assess sources of danger in engineering and mining. In: Natau, O., Fecker, E., Pimentel, E., (Eds.), Geotechnical measurements and modeling. A.A. Balkema.

Topp, G.C., Davis, J., and Annan, A., 1980. Electromagnetic determination of soil water content. Water Resources Research 16 (3), 574-582.

Toshioka, T., Tsichida, T., Sasahara, K., 1995. Application of GPR to detecting and mapping cracks in rock slopes. Journal of Applied Geophysics 33, 119-124.

Twiss, R.J., Moores, E.M., 1992. Structural geology. W.H.Freeman and Company, New York.

Unterberger, R.R., 1978. Radar propagation in rock salt. Geophysical Prospecting 26 (2), 312-328.

Van der Kruk, J., Wapenaar, C.P.A., Fokkema, J.T. and Van den Berg, P.M., 2003. Improved Three-dimensional Image Reconstruction Technique for Multi-Component Ground Penetrating Radar Data. Subsurface Sensing Technologies and Applications 4 (1), 61-99.

Ward, S.H., and Hohmann, G.W., 1987. Electromagnetic theory for geophysical applications, *in* Electromagnetic methods in applied geophysics, Eds. Nabighian, M. N., *Soc. Expl. Geophys*, Vol. 1, 131-311.

Warren, J.K., 1989. Evaporite sedimentology: importance in hydrocarbon accumulation. Prentice Hall, New Jersey.

Warren, J.K., 2010. Evaporites through time: Tectonic, climatic and eustatic controls in marine and nonmarine deposits. Earth-Science Reviews 98, 217-268.

Whyatt, J.K., Varley, F.D., 2008. Catastrophic failures of underground evaporite mines. Proceedings of the 27th International Conference on Ground Control in Mining, July 29 - July 31, 2008, Morgantown, West Virginia. Peng SS, Mark C, Finfinger GL, Tadolini SC, Khair AW, Heasley KA, Luo-Y, eds., Morgantown, WV, West Virginia University, 113-122.

Yaramanci, U., 2000. Geoelectric exploration and monitoring in rock salt for the safety assessment of underground waste disposal sites. Journal of Applied Geophysics 44, 181-196.

Yechlakov, Y.A., Morozov, G.G., 2006. Stratigraphy. *in* Mineral resources of Perm kray. Ed. A.I. Kudryashov. Knizhnaya ploschad, Perm, Russia, 49-63 (in Russian).

Young, R.A., Deng, Z., Marfurt, K.J., Nissen, S.E., 1997. 3-D deep filtering and coherence applied to GPR data: A study. The Leading Edge 16 (6), 921-928.

Zeng, X., McMechan, G. A., Xu T., 2000. Synthesis of amplitude-versus-offset variations in ground-penetrating radar data. Geophysics, 65 (1), 113-125.

Zipf Jr., R.K., Swanson, P., 1999. Description of a large catastrophic failure in a southwestern Wyoming Trona Mine. Rock Mechanics for Industry, Ed. Amadei, Kranz, Scott and Smeallie, Balkema, Rotterdam, 293-298.

Zonenshain, L.P., Kuzmin, M.I., Natapov, L.M., 1990. Geology of the USSR: a Plate-

Tectonic Synthesis. American Geophysical Union, Washington, DC.

# VITA

Oleg Nikolaievich Kovin was born in Perm, Russia, on March 14, 1954. In 1971, he graduated from Middle School 119, in Perm, Russia. In 1971, Mr. Kovin entered the Geological Department of the Perm State University, Russia. He received the degree of Specialist (Master of Science) in Geology and Geophysics in 1976.

In 1976, Mr. Kovin joined the Tashkent Geology, Inc., Uzbekistan as engineer-geophysicist. From 1978 to 1980, he was involved in the research projects in the Institute PermOil, Perm, Russia. In 1980, Mr. Kovin obtained a position as research scientist in the Kama Branch of All Union Oil and Gas Geological Institute, Perm, Russia. From 1982 to 1988, he worked as senior research scientist in the Perm Polytechnic Institute (currently Perm Technical University), Perm, Russia. In 1988, Mr. Kovin joined the Mining Institute of Ural Branch of the Russian Academy of Sciences as research scientist. In 2004, he enrolled in the PhD program in the Department of Geological Sciences and Engineering of Missouri University of Science and Technology (former University of Missouri-Rolla).

Mr. Kovin has a number published conference and journal papers, some of which are included in the reference list of this research. He was involved in various research and geophysical survey projects including ground penetrating radar, seismic, electromagnetic, resistivity, and hydroacoustic methods.

He has been a member of the Society of Exploration Geophysicists (SEG) since 2006 and a member of the Environmental and Engineering Geophysical Society (EEGS) since 2008.

CPSIA information can be obtained
at www.ICGtesting.com
Printed in the USA
LVIC011518080213
319330LV00011B

*9 7 8 1 2 4 9 0 3 6 0 6 7*